Apress Pocket Guides

Apress Pocket Guides present concise summaries of cutting-edge developments and working practices throughout the tech industry. Shorter in length, books in this series aims to deliver quick-to-read guides that are easy to absorb, perfect for the time-poor professional.

This series covers the full spectrum of topics relevant to the modern industry, from security, AI, machine learning, cloud computing, web development, product design, to programming techniques and business topics too.

Typical topics might include:

- A concise guide to a particular topic, method, function or Framework
- Professional best practices and industry trends
- A snapshot of a hot or emerging topic
- Industry case studies
- Concise presentations of core concepts suited for students and those interested in entering the tech industry
- Short reference guides outlining 'need-to-know' concepts and practices.

More information about this series at https://link.springer.com/bookseries/17385.

The Data Flow Map

A Practical Guide to Clear and Creative Analytics in Any Data Environment

Nick Ryberg

Apress®

The Data Flow Map: A Practical Guide to Clear and Creative Analytics in Any Data Environment

Nick Ryberg
Lake Elmo, MN, USA

ISBN-13 (pbk): 979-8-8688-1882-0 ISBN-13 (electronic): 979-8-8688-1883-7
https://doi.org/10.1007/979-8-8688-1883-7

Copyright © 2025 by Nick Ryberg

This work is subject to copyright. All rights are reserved by the Publisher, whether the whole or part of the material is concerned, specifically the rights of translation, reprinting, reuse of illustrations, recitation, broadcasting, reproduction on microfilms or in any other physical way, and transmission or information storage and retrieval, electronic adaptation, computer software, or by similar or dissimilar methodology now known or hereafter developed.

Trademarked names, logos, and images may appear in this book. Rather than use a trademark symbol with every occurrence of a trademarked name, logo, or image we use the names, logos, and images only in an editorial fashion and to the benefit of the trademark owner, with no intention of infringement of the trademark.

The use in this publication of trade names, trademarks, service marks, and similar terms, even if they are not identified as such, is not to be taken as an expression of opinion as to whether or not they are subject to proprietary rights.

While the advice and information in this book are believed to be true and accurate at the date of publication, neither the authors nor the editors nor the publisher can accept any legal responsibility for any errors or omissions that may be made. The publisher makes no warranty, express or implied, with respect to the material contained herein.

> Managing Director, Apress Media LLC: Welmoed Spahr
> Acquisitions Editor: Shaul Elson
> Development Editor: Laura Berendson
> Coordinating Editor: Gryffin Winkler

Cover designed by eStudioCalamar

Distributed to the book trade worldwide by Springer Science+Business Media New York, 1 New York Plaza, New York, NY 10004. Phone 1-800-SPRINGER, fax (201) 348-4505, e-mail orders-ny@springer-sbm.com, or visit www.springeronline.com. Apress Media, LLC is a Delaware LLC and the sole member (owner) is Springer Science + Business Media Finance Inc (SSBM Finance Inc). SSBM Finance Inc is a **Delaware** corporation.

For information on translations, please e-mail booktranslations@springernature.com; for reprint, paperback, or audio rights, please e-mail bookpermissions@springernature.com.

Apress titles may be purchased in bulk for academic, corporate, or promotional use. eBook versions and licenses are also available for most titles. For more information, reference our Print and eBook Bulk Sales web page at http://www.apress.com/bulk-sales.

Any source code or other supplementary material referenced by the author in this book can be found here: https://www.apress.com/gp/services/source-code.

If disposing of this product, please recycle the paper

This book is dedicated to Kathryn, the love and light of my life.

Table of Contents

About the Author .... **xiii**

About the Technical Reviewer .... **xv**

Acknowledgments .... **xvii**

Introduction .... **xix**

Chapter 1: Introduction **1**
   Where Can the DFM Framework Help? 4
   Where Is the DFM Not Useful? 4
   Who Will Benefit from the Framework? 4
      Analyst 5
      Data Scientist 5
      Engineer 5
   Summary 6

Chapter 2: Framework Overview **7**
   Modes 7
   Stages 8
   Actions 9
   Breaking It Down 10
      Source 10
      Focus 11
      Build 11

TABLE OF CONTENTS

First Walkthrough 12
    Source 12
    Focus 13
    Build 13
    Result 14
Summary 15

Chapter 3: Data Flow Map Deep Dive 17

Annotations 18
Source 19
    O Get 20
    + Mix 22
    X Fix 24
    Source: Start at the beginning 24
Focus 25
    V Cut 27
    > Slice 28
    Row Limit 29
    Conditional Value 30
    Lists 31
    Nulls/Blanks 31
    Sampling 32
    \ Sort 33
    Focus: Get to the point 34
Build 34
    ] Box 35
    # Size 36
    = Ship 39
    Save to File 40

TABLE OF CONTENTS

Build: Shipping a Conclusion 41
Tag 42
    What's in a Tag? 42
    Examples 43
    Keyword Conflicts 43
Summary 44

Chapter 4: Examples – Files 47
Files 47
Source 48
    o Get 48
    + Mix 49
    x Fix 51
Focus 57
    v Cut 57
    > Slice 58
    \ Sort 60
Build 62
    ] Box 66
    # Size 70
    = Ship 74
Summary 75

Chapter 5: Examples – Databases – SQL 77
Databases 77
    Source 78
    Focus 83
    Build 88
Summary 93

TABLE OF CONTENTS

Chapter 6: Examples – Python 97
Python Code 97
Source 98
o Get 98
+ Mix 100
x Fix 102
Focus 106
v Cut 107
> Slice 108
\ Sort 109
Build 112
] Box 112
Size 113
= Ship 115
Summary 116

Chapter 7: Cloud API 119
@ Land the Data 121
Source 121
@ Pull Multiple Stations 123
Source 123
Focus 125
Build 127
@ Process JSON to CSV 129
Source 129
@ Top Five States 130
Source 131
Focus 131

TABLE OF CONTENTS

    Build 132

    Summary 135

Chapter 8: Platforms 137

    Source 138

        O Get 139

        X Fix 140

        X Mix 141

    Focus 144

        V Cut 144

        > Slice 146

        \ Sort 147

    Build 149

        ] Box 149

        # Size 149

        = Ship 150

    Summary 153

Chapter 9: Examples – Pipelines 157

    Pipelines 157

    Source 159

        O Get 159

        + Mix 161

        X Fix 163

    Focus 165

        V Cut 165

        > Slice 167

        \ Sort 168

TABLE OF CONTENTS

Build 169
    ] Box 171
    # Size 172
    = Ship 174
Summary 179

Chapter 10: Analog – Pen and Paper 181

Example Data 182
French Bread – The Data Flow Map 184
    @ Proofing Yeast 184
    @ Create Dough 184
    @ Rising Dough 185
    @ Shaping 186
    @ Baking 187
Summary 188

Appendix A: Sample Data Sourcing 193

About the Author

Nick Ryberg has built analytics across various platforms, from Microsoft Excel and Access to more advanced systems like Postgres, Hadoop, and Spark SQL. Whether he's working on personal computers, Linux servers, mainframes, or even a Raspberry Pi, Nick thrives with a keyboard and a table or two of data.

As tools improve, becoming more user-friendly and capable of handling larger datasets, Nick has noticed that how we think and share our processes hasn't changed much. At best, it's a messy whiteboard with bubbles and arrows; at worst, it's raw code left behind by a developer who left years ago.

Throughout Nick's career, he has focused on solving challenging analytic problems using these tools. The most complex problems he encounters aren't related to sourcing, cleaning data, or mastering specific tools. Instead, the hardest parts involve thinking differently about solutions, sharing and brainstorming ideas, switching platforms, and documenting processes for future users.

About the Technical Reviewer

Aniket Sundriyal has over 14 years of experience as a data science and analytics leader, specializing in machine learning, customer lifecycle strategies, and large-scale reporting solutions. He is passionate about advancing the field and has served as a technical reviewer for leading AI and data science conferences. Aniket's reviews focus on real-world applicability, emphasizing measurable business impact, modeling rigor, and scalable system design. He is a Fellow of the British Computer Society (BCS), the Institution of Electronics and Telecommunication Engineers (IETE), and the Institute of Analytics (IoA).

Acknowledgments

I am deeply grateful to my wife, Kathryn Kerwin-Ryberg, for guiding a collection of wildly different ideas into a unified whole, offering brilliant and funny ideas, and giving me the space and time to bring this book to life. I want to thank the team at Springer Nature/Apress Publishing for recognizing the value of a crazy idea and helping to bring it into book form.

Introduction

Life, your work, and the world **happen**. **Data** tracks what *happens* with events and measures motion and the shape of things. **Analytics** source *data*, focus on the important, and build new conclusions, changing your *life, work, and world* for the better. Every analysis follows a pattern, often repeated several times in a path to success:

Source: Get raw data, mix it with other contexts, and fix it for later analysis.

Focus: Select specific columns, the right rows, and in the best order.

Build: Categorize and size your data into meaningful chunks.

Using the data and tools you already have, this book will help you develop analytics more thoughtfully, effectively communicate the process with your team and key partners, and ultimately deliver better analytics that drive change in the right direction.

CHAPTER 1

Introduction

Over the years, I have used a wide range of analytic tools across various data sources. From the smallest spreadsheet pivot tables to gigabyte-sized sales analytics using high-performance computing, I've seen a range of challenges to gain a deeper understanding of the world and make better decisions with data.

While each tool and data source opened possibilities for better analytics, company-driving incentives, and goal-reach possibilities, there was always a key weakness I struggled with at the end of the day.

Each final analysis was shrouded in technical details that made it difficult to think clearly, be creative, and share the process of how I arrived at the solution with my team members, supporting engineers, and client users. Whenever I returned to a project I worked on more than a day ago, I struggled with the bigger picture of the solution.

The data wasn't my problem – modern hardware has made it easier and easier for me to dig into vast amounts of information, returning results in minutes, if not seconds, in volumes unimaginable even a few years ago.

The tooling wasn't my problem – the broader ecosystem of analytic tools has exploded with more user-friendly approaches to dig into the mountains of data. And if I may be so modest, I've gotten pretty good at writing SQL in an old-school database.

CHAPTER 1 INTRODUCTION

The problem was that I kept getting lost in the implementation and forgetting the bigger picture of why I was building analytics in the first place – answering tricky questions and opening new lines of analytic investigation. Three key blocks kept popping up over and over:

Creativity: I created brittle code and confusing, often circular analytics, stifling my ability to think outside the box. I didn't dare touch legacy production code. The whole thing would come crashing down and I needed this report yesterday! I'd leave it the way it was and duct tape worse solutions on top of the code heap.

Communication: I onboarded new team members and collaborators who sincerely wished they could read my mind. My analytics were robust, earth-shaking solutions, but how did I solve the problem in the first place? Where did the data come from? The most common phrase when sharing my solutions was, "I should clean this bit up first," followed by, "Sorry, the documentation just needs a little love," and then a sheepish, "Just look at the code." Unless my reader held my exact level of expertise (good or bad), reading the code wouldn't help them.

Clarity: As my analyses grew, I began to see multi-step analytic processes as the norm, rather than the exception, especially when I had to cross platform boundaries and language barriers. Each of my analytical steps had domain-specific issues, making it challenging to grasp the overall picture and identify potential areas for improvement.

How and where the dots connected was often the most significant pain point. It was impossible to keep everything in my head and ensure that I was answering the original question correctly and efficiently.

At first, I thought the challenges were related to this tool and that data, but as I changed jobs and companies, seeing new tooling and data sources, the problems persisted. When I complained that this vendor's tool and that department's database were driving me crazy, a brilliant friend pointed out that it's not about the tools, and it's not how the tables are organized. How I was framing up the problems in my head before I took the first step was the core of the problem.

I realized that, regardless of the tooling and data sourcing, common patterns emerged with each analysis. It didn't matter if I was pivoting a CSV file in Excel into a bar chart in a few seconds or taking hours to process billions of Kafka records using Python and Spark; the same key ideas kept coming up to the surface.

I took notes on those key ideas and iterated through multiple notation systems and Frameworks. My goal was to capture an analysis from beginning to end in a way that captured everything at a level just above the actual code and data, and just below a problem statement.

As my Framework started to take shape, I kept a few key rules in mind: First, the Framework had to be easy to understand and remember – it's not helpful if a Framework is more verbose than the code it describes. Second, I wanted the Framework components to fit into three or four primary motions. I was sure that all the things we do in an analytical process boil down to just a few key ideas. And finally, it had to be universal. I didn't want to create a Framework that was only applicable to SQL and didn't fit Excel or Python scripts.

CHAPTER 1   INTRODUCTION

The work over the last several years has boiled down to the Data Flow Map Framework. It's simple and universal, and it has revolutionized my creativity, communication, and clarity in my analytical projects. I hope it is just as helpful for you and that you find yourself thinking in terms of modes and actions.

Where Can the DFM Framework Help?

The Framework helps structure any analysis clearly and effectively, whether tabulating attendance on pencil or paper, tallying daily sales, or creating the next big machine learning model.

Where Is the DFM Not Useful?

This is not the tool or the data. There is no software to install. This involves rethinking how to approach analytics, utilizing the tools and data at hand to create new solutions and tell that story.

Who Will Benefit from the Framework?

This Framework is for anyone who works with data for more than a few days, particularly those who analyze the data, test their hypotheses, or apply engineering principles to build data foundations. Where users cross boundaries, using multiple tools and think like an analyst, scientist, and engineer in their workflow, the DFM is beneficial for creating a common language to boost understanding and efficiency

Analyst

Whether by title or function, millions of people gather, clean, summarize, and extract the key information for a specific audience and goal. Their technical skills vary wildly from novices fresh from college to seasoned power users that the team can't live without. Their key challenge is sharing work, which is limited by their approach to solutions and their inability to communicate processes effectively. This makes it particularly challenging to onboard new users as responsibilities change, conveying their story to non-technical stakeholders, and obtaining usable feedback.

Data Scientist

Data scientists need to test and learn from data, often using Python or R, and they often curse the time spent cleaning data instead of creating business-changing insights with complex statistical models.

Processing data can take hours or even days. Gaining a clearer understanding of their process would help prevent future misunderstandings with clients and partner engineers, paving the way for better and faster solutions. Their main challenge is explaining how they arrived at their results so that clients can understand the benefits and limitations of the complex work.

Engineer

This group is essential for any company with analysts and data scientists. They access large data sources using SQL and other powerful tools, creating a data foundation that allows analysts and scientists to generate game-changing insights. The data volume can be immense, and properly aligning pipelines can be a significant challenge. Engineers usually develop complex, lengthy processes for moving, cleaning, and preparing

CHAPTER 1   INTRODUCTION

data. Their main challenge is communicating effectively with others to ensure understanding and agreement on accuracy, consistency, and reliability.

Summary

The Data Flow Map Framework enhances your ability to manage complex analytics, clarify your approach, and even elevate it to the next level. At their core, analytics, data science, and data pipelines focus on sourcing data, emphasizing the crucial aspects, and constructing new structures from the data.

Next, I will provide a high-level overview of the Framework, followed by a chapter that provides an in-depth exploration. The remaining chapters will bring the Framework to life through various tools and platforms using different sample datasets.

CHAPTER 2

Framework Overview

In the introduction, I explored the daily challenges and opportunities analysts face. Starting with this chapter, I'll offer the Data Flow Map as a solution that will help connect tools, data, and analysts' brains more effectively.

Think of this as the high-speed run-through before the complete breakout in the next chapter.

Modes

Every analysis, whether a back-of-the-napkin estimate, a comprehensive financial spreadsheet model, or a Python-driven model, progresses through three main **modes**:

1. **Source**: Gather raw data, mix it with other sources, and prep it.
2. **Focus**: Select the correct data, and sort top to bottom.
3. **Build**: Create new patterns with groups, measures, and a final result.

The Data Flow Map Framework follows these modes in a straightforward, visual, and user-friendly way. The modes will typically follow the order shown in Figure 2-1.

CHAPTER 2   FRAMEWORK OVERVIEW

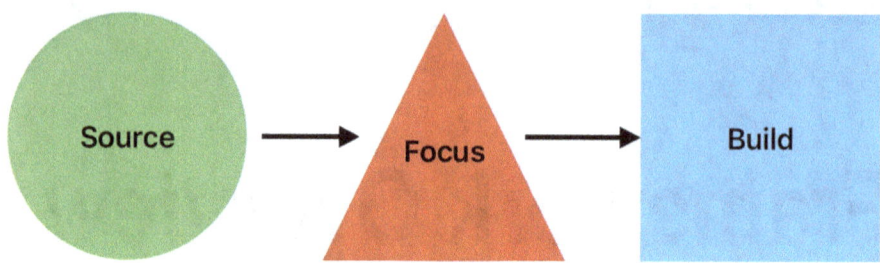

Figure 2-1. Data Flow Map Framework Foundation Modes

Stages

It's rare for an analysis to rely on only one source, one focus, and one build. Modes can be stacked in stages, creating more complex analyses. For example, the analysis might source data from different locations, which could require multiple stages as the analysis builds a bigger story, as shown in Figure 2-2.

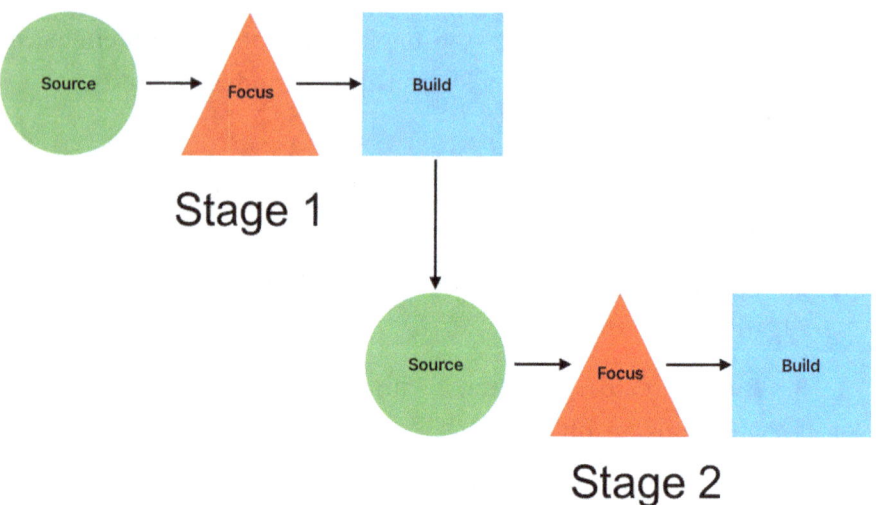

Figure 2-2. Stacked Data Flow Maps

8

CHAPTER 2 FRAMEWORK OVERVIEW

Each of these stages is assigned a tag. This simplifies understanding and documenting the much larger multi-stage analytics for future reference. In the example above, they are called "Stage 1" and "Stage 2"; however, I will discuss this in more detail in the next chapter, explaining how breaking your analytics into discrete, tagged parts can enhance the entire process.

For now, let's keep digging deeper into each mode with specific actions.

Actions

Each mode (Source, Focus, Build) is broken down into three smaller actions, each with a unique name and symbol, for a total of nine actions overall, see Figure 2-3.

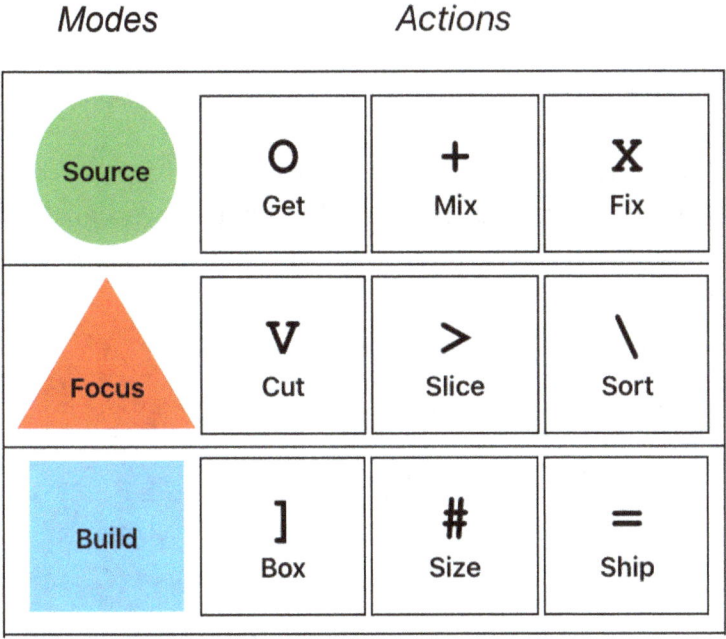

Figure 2-3. *Data Flow Map Modes and Actions*

These actions are the most common individual verbs that can be applied to data to help build an analytic process. I've intentionally chosen very short, punchy words with clear symbols that are available on any keyboard or can be easily handwritten.

The modes and their actions have been developed through the evaluation of thousands of analyses to provide the simplest yet most comprehensive way to describe any process clearly and effectively.

An entire analytic map might consist of just two actions in two modes, or it might involve multiple stages with a variety of actions, describing a complex path from raw data to the final report.

Breaking It Down

In this section, I will walk through each mode and action individually, explaining their purpose and how they fit together.

Source

The first mode is about gathering data from one or more sources and making it usable for analysis. This is the beginning of every analysis, or it can be repeated through multiple stages as the data approaches a final result.

See Table 2-1 for a listing of the three Source Actions.

Table 2-1. Source Actions

Symbol	Action	Purpose
o	Get	Source a single location of data
+	Mix	Incorporate other sources
x	Fix	Format, reorder or extract useful data

CHAPTER 2   FRAMEWORK OVERVIEW

 Focus

In Focus mode, concentrate on what matters and remove what doesn't. Focus is about cutting out distractions and highlighting the most important information in your analysis. Understanding these choices and making them easy to grasp is often the biggest challenge when communicating results to users, blocked out in Table 2-2.

Table 2-2. Focus Action

Symbol	Action	Purpose
V	Cut	Select columns that matter
>	Slice	Select rows that are critical
\	Sort	Order most to least important

 Build

Finally, in Table 2-3, shake up your analyses and make a point. In the first two modes, I discussed locating data, making it useful, and concentrating only on what is essential. This mode involves building new towers of information on a solid foundation by grouping, sizing, and delivering the final result. Table 2-3 lists the primary Build actions.

Table 2-3. Build Actions

Symbol	Action	Purpose
]	Box	Summarizing the data into buckets
#	Size	Measuring the size of the buckets
=	Ship	Deliver final results

11

CHAPTER 2   FRAMEWORK OVERVIEW

First Walkthrough

The next chapter examines each mode and the actions that comprise the Data Flow Map Framework. To solidify this Framework, I would like to start with a simple example.

I believe it's important to shop locally for coffee. Since I have a comprehensive personal history of purchasing coffee at shops both large and small, I'd like to know if I walk the walk and drink the local drink, or if I'm caving to temptation with the national chains.

This walkthrough uses my history of coffee purchasing habits, which is explained in more detail in Appendix A.

I want to build an analysis of my coffee-drinking patterns and favorite stores. I have two data sources:

> **Coffee Purchases**: date, store, and amount

> **Location Category**: the size of the company that owns the coffee shop – local, regional, or national

The analysis will cover all three modes, each tagged with a simple name. This walkthrough will focus only on the modes and actions. In Chapter 5, I'll start incorporating the code used to process the data.

 **Source**

This is where the analysis starts – sourcing the raw data and preparing to create something amazing. In Table 2-4, I'll **GET** the primary purchase history and **MIX** in the location categories. I'll also **FIX** some of the data to make it more useful later.

CHAPTER 2   FRAMEWORK OVERVIEW

Table 2-4. *Sourcing Coffee Data*

@	Action	Fetch Sales
O	Get	Gather Coffee Purchase History
+	Mix	Add in store category based on payee
X	Fix	Create a purchase year from the purchase date and round amount to whole dollars

 Focus

Focusing is about limits – how rapidly can we narrow down the large source dataset to just the columns and rows that help answer my questions? In Table 2-5, I'll **CUT** and keep the valuable columns, **SLICE** the results to keep local locations, and **SORT** by purchase year.

Table 2-5. *Focus Actions*

@	Action	Sharpen the Picture
V	Cut	Focus on store category, purchase year, and whole dollar amount
>	Slice	Select "local" Category
\	Sort	Sort by the purchase year ascending

 Build

Finally, I want to **BUILD** something valuable from the **SOURCED** and **FOCUSED** data. In Table 2-6, I'll begin creating the final product by summarizing the data by location type (**BOX**), totaling the sales and counts (**SIZE**), and making an output file (**SHIP**).

13

CHAPTER 2 FRAMEWORK OVERVIEW

Table 2-6. Summarize the Results

@	Action	Group Location
]	Box	Summarize according to location Type
#	Size	Sum total sales and counts
=	Ship	Create a final CSV file summarized values

Result

Figure 2-4 shows that my purchasing habits at "Local" coffee shops have varied considerably over several years. Coffee buying peaked in 2022 and has since declined.

Year	Category	Count	Sum Amount
2014	Local	14	64
2015	Local	16	70
2016	Local	12	141
2017	Local	5	48
2018	Local	4	104
2019	Local	6	89
2020	Local	3	32
2021	Local	5	73
2022	Local	44	488
2023	Local	4	48

Figure 2-4. Coffee Purchases by Year for Local Shops

Based on this data, I could take the analysis in various directions: What happens to purchasing behavior in other location categories? Are annual amounts consistent from year to year?

Because I've built a Data Flow Map around the analysis, I now have a simpler and clearer model of where the data originated (SOURCE), the choices I made to refine it (FOCUS), and my approach for summarizing it (BUILD).

With a clear source, focus, and build, I have the flexibility and courage to take it in multiple directions without losing sight of the bigger picture. I can uncover new and interesting patterns. If I switch to a different analytical tool, such as Excel for graphing, it would be relatively easy to make the jump. Finally, explaining the overall process to anyone and having a conversation about choices and opportunities for the analysis is much easier.

Summary

The Data Flow Map Framework provides a higher-level perspective on foundational analytics. No matter how simple or complex, every analysis follows a consistent pattern: Sourcing data, Focusing on critical points, and Building results with impact.

- **Source**: Get data, mix it with other data, and fix as necessary.
- **Focus**: Cut the columns and slice rows, sorting the most important to the top.
- **Build**: Boxing the data into groups, sizing, and shipping results.

A more complex analysis can be broken down into the larger chunks using stages to help separate the sequences. As multiple stages are built and documented, consolidating data and behavior together is easier and more transparent.

CHAPTER 2 FRAMEWORK OVERVIEW

The execution of your analysis depends on your analytical toolbox, skills, and the size of your datasets. Your approach to process can be enhanced by applying the same Framework patterns using Excel, SQL in a Database, or Python with Pandas, moving from raw data to insights, visuals, and new opportunities that can transform your world.

The next chapter of this guide will explore the Framework more thoroughly, focusing on the details of each possible action. Finally, the rest of the book will include examples of analytics using various tools, platforms, and solutions. While each chapter will feature different data and technical approaches, the overall Framework of source, focus, and build will consistently follow the nine actions.

CHAPTER 3

Data Flow Map Deep Dive

In this chapter, I'll take the high-level overview from the previous chapter and discuss each segment in more depth (see Figure 3-1). I'll explain the logic behind each action and identify a theme that runs throughout the entire process.

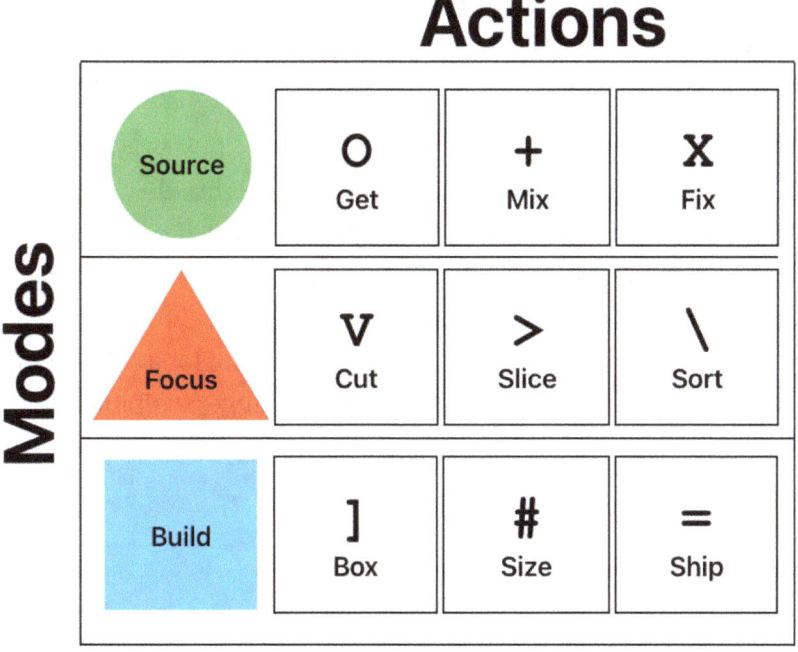

Figure 3-1. Data Flow Map Elements

CHAPTER 3   DATA FLOW MAP DEEP DIVE

Annotations

Throughout the book, I'll be using a table format to organize the different actions into stages. In this deep dive, the tables will show a simple piece of work, labeled with a tag to help identify it.

As I continue through the rest of the book, these tables will begin to resemble actual stages in an analysis, and you'll see all of the actions involved. For now, however, they will be quite fragmented and limited to just the mode and action I am focusing on. Table 3-1 shows a simple example.

Table 3-1. Sketching out the big steps

@	Choices Matter	Pick only the variables that change the world
o	Get	Coffee Data
V	Date	When did the purchase happen?
V	Amount	How much was it worth?

The "@" indicates the stage tag – in this case, "Choices Matter." Where suitable, I will include explanatory details on the right side.

The first column shows the symbol used for a specific action, the second displays the action name, and the third provides a more conversational description of what is happening.

When stages begin to connect, it will be common for a subsequent stage to refer to the previous stage's result as a source; see Tables 3-2 and 3-3.

Table 3-2. *Stage One – Source the Data*

@	Choices Matter	Pick only the variables that change the world
o	Get	Coffee Data
V	Date	When did the purchase happen?
V	Amount	How much was it worth?
=	Ship	Hold the Data for next stage

Table 3-3. *Stage Two – Make Sense of It*

@	Summarize	Simplify the data and measure results
o	Get	Use Choices Matter
]	Box	Group by year
#	Size	Total up purchase amounts
=	Ship	Create a simple bar graph

The goals throughout the Data Flow Map Framework are to make it as easy as possible to think about complex analytics, create brand new results, and tell the story of the process in a way that anyone can understand. By breaking a multi-stage analysis into smaller stage chunks, it will be easier to reason about the bigger process. Let's start diving into the actions, starting with the Source mode.

Source

Data Flow Maps represent motion – moving ideas from one place to another. Sourcing data begins with a Get for the primary source, sometimes a Mix for context, and almost always includes one or more Fixes to shape the data into a usable format. See Figure 3-2 for the key actions.

CHAPTER 3 DATA FLOW MAP DEEP DIVE

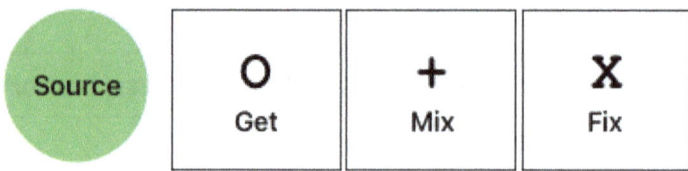

Figure 3-2. Source Mode's Three Actions

O Get

The first step out the door is finding a set of facts. Amid the swirling snowstorm of data, a tiny, bright pinprick of light shines in your eye. It may have been given to you, or it could be data you've captured by hand, gradually constructing a picture.

Every analysis starts with a primary set of facts. You will likely include other contextual information, but this first "GET" pulls facts that form the backbone of the analysis – they could be sales, visits, trips, inventory, people, money, or resources. This step can be thorough – initially, you identify a specific resource by its complete name and possibly provide additional information on how others can locate it.

The GET statement can be a simple line that describes a location (database, file server, or website) and a table name.

Format

The primary source of an analysis falls into one of four major categories: files, databases, in-memory, and streaming.

Files

CSV (Comma-Separated Values) files are the most common file format worldwide. Every analytic tool available supports this format natively. While there may be issues with column types, naming, and validation, rest assured that any tool you use will offer techniques to make imports as straightforward as possible.

Excel spreadsheets are the second most common format, which, unlike CSV, is less widely readable. It's rare for a business user not to have Microsoft Excel installed on their computer and ready to process this format. Most spreadsheet-like tools, such as Google Sheets, can make importing data in this format easier.

Finally, there are binary file formats like Parquet, often used for data science or transferring data between systems.

Databases

A database is a system that organizes data in table form and can be accessed locally or remotely, usually using SQL. There are different types of database servers, both local and remote, with SQLite and PostgreSQL being the most common. If you work in data engineering, you will likely have at least one primary database, possibly several. Most analysts who extract data from a database use SQL to retrieve it.

In-Memory

After the first step, you can have data in memory. It could be a data frame, a common table expression, or something in your computer's clipboard.

If the analysis involves multiple stages, temporarily store the data in memory as a holding spot, then refer to it in later steps using tags. Most analytic tools allow you to save data temporarily for future use.

Streaming

The data can originate from a website, an API, or a continuous source, typically seen when the data source is constantly changing, such as sales in a store or a temperature monitor. These flows are usually processed by engineered pipelines that handle the data for later analysis, storing it in files or a database.

+ Mix

One of the qualities that define us as human beings is our ability to connect context with our current situation so that we can see the bigger picture and make better decisions. For example, we start the day by reviewing our schedule, considering where we need to go, and checking the weather forecast. This might alter our plans for commuting, clothing choices, and goals for completing tasks.

Good analytical decisions are made by combining multiple sources of information to create a clearer picture of the entire world.

Context

A data table is usually two-dimensional, composed of rows and columns. You can perform interesting tasks in Build mode to extract and summarize information. Incorporate the surrounding data that describes your situation; this allows you to be more creative in your analytics, leading to new solutions and opportunities.

For decades, Excel has enabled us to join two tables using the VLOOKUP function. This process is delicate and prone to errors, but when successful, it allows you to merge two sets of information. Even without external data, you can at least think beyond the limitations of the tables into something broader and significant.

Analysts must continually optimize for data density and speed since we have limited time and resources. Capturing all associated data for an event can generate terabytes of information, so we depend on shortcodes. For example, instead of spelling out the full city or neighborhood name, we use five-digit zip codes to geographically label locations in the United States. But what if we want to analyze data at the state level? Incorporating the context of zip codes will enhance clarity and make the data easier to interpret.

CHAPTER 3   DATA FLOW MAP DEEP DIVE

If your data is filled with acronyms and shortcodes, especially in legacy systems, you can include an explanatory table that maps the values to something more understandable and readable:

- 55023: Minnesota
- 50047: Iowa
- 99501: Alaska

By creating mapping tables, you can make your analysis much more approachable for the average reader. Adding extra color and style to your content will make it more engaging and accessible to other readers.

Join Types

Depending on your analytic tool, you may have several choices for how the data are joined together.

- **Inner**: The default choice for most joins, where the values match, they're brought together.
- **Left/Right Join**: Include all values from the left or right side of the join, along with the matching values from the other side.
- **Outer**: Connect data regardless of whether a perfect join exists. This can help identify gaps.
- **Anti-Join**: Less common but very useful to only select rows where the two tables don't meet.
- **Cross**: The least common join used for joining every row of one table to every row of a second, resulting in often extensive results.

x Fix

Data sources are imperfect, and fixing them is what makes you excel at your work. There will be rough edges, fields that don't align properly, or areas where your analytical tool needs refinement. Very few things in this world are perfect without flaws, stains, or minor room for improvement, except the love of your life.

Some repairs are needed, and tears in the data fabric must be stitched. Lastly, a key aspect of Boxing depends on categories, and "Fixed" data can provide that grouping level to otherwise continuous data. For example, if you're working with dates and want to Box by year, and your source data is at the daily level, you'll need to adjust the date to the annual level.

The most common data-fixing patterns are date manipulation, type casting, and relabeling. The last is the easiest. If you have columns with abbreviated titles and tables with hard-to-read short names, expand them into longer phrases so you can tell the difference between "`Sls_Curr`" and "`Sls_Curr_LY_Pct`." Conversely, if you need to label fields with their names, you can quickly relabel a table with a clear name to ensure you understand the distinction between mixed-in data sources.

Source: Start at the beginning

In some cases, for privacy reasons, you should mask data. For example, you need to handle this if you have a potentially personal and hackable Social Security number or other form of identification. Fixing it by hashing or simply redacting will go a long way toward safer data.

To summarize the Source mode, every analysis will have a primary data source. It could be a file, a database, or something on the web. Most analytics that tell a larger story incorporate more context by mixing in other data sources. Finally, almost every dataset requires modifications or repairs to make it usable for your goals.

Using the table format for documentation, the Source Mode would look like Table 3-4.

Table 3-4. *Source Mode Summary*

@	Landing Data	Sourcing the right data and prepping
o	Get	Coffee Data CSV File
+	Mix	Add in Store Categories
X	Fix	Modify daily dates to just year

Once you have identified a Source, Mixed context, and Fixed problems, it's time to get a lock on what's essential – the Focus mode!

Focus

We live in a golden age of information. With user-friendly devices at our fingertips and in our pockets, we can find, define, or explain just about anything in a few clicks. To meet your analytic goals, you must quickly make sense of one or two of these data nuggets, especially those with the sharpest teeth or potential for dinner. Whatever your aim, analysis will always boil down to a limited set of facts. Focus is the process that helps you get to that point as quickly as possible.

In the past, computers faced major limitations. Analyzing data was a programming challenge, and ensuring a program responded correctly was a lengthy and complex process. You carefully chose which data to display because you didn't have all day to wait for a program to compile and produce results. As a result, you had to decide what you needed before doing an analysis, which limited creativity.

Our computers now have the advantage of vast amounts of memory and almost unlimited disk space, allowing them to gather billions of rows of data across hundreds of columns. Yet, our brains remain stuck

CHAPTER 3   DATA FLOW MAP DEEP DIVE

in caveman mode, focused on what will keep us safe and support our families. The most limited resource is attention, and your task is to ruthlessly narrow down the data to only what has the greatest impact.

Focusing comes in three flavors – see Figure 3-3.

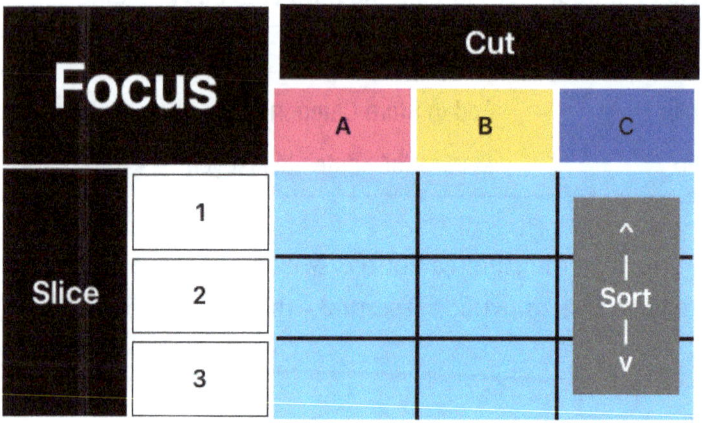

Figure 3-3. Focus by Actions – Cut, Slice, and Sort

- **Cut**: A vertical slice of the data, selecting the columns or fields that matter most. This is the "which" question – in this graphic, it's the A, B, and C columns that show which dimensions and values are important.

- **Slice**: A horizontal slice that selects only the rows of data matching filter conditions. This addresses the "what" question, often expressed as a "WHERE," represented here by the 1, 2, and 3 horizontals, and brings together the selected values to tell the story effectively.

- **Sort**: Arrange data so that the most important, whether small or large, appears at the top, by ordering rows based on selected columns.

Throughout the "Focus" section, I'll use this visual to highlight how the actions fit together.

CHAPTER 3 DATA FLOW MAP DEEP DIVE

V Cut

A cut removes one or more columns, leaving only the variables you need. In Figure 3-4, I'm keeping columns A and C, dropping B.

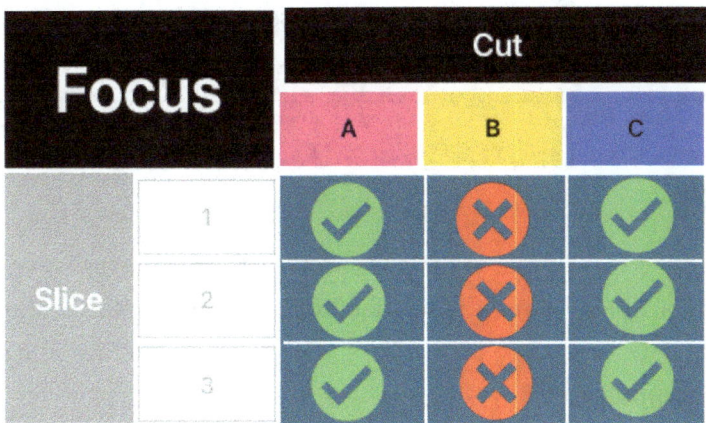

Figure 3-4. Cut

Everything you do in the Sourcing phase allows you to preserve what matters selectively. Cutting data involves choosing "which" parts of the data – external chunks of variables, fields, and columns.

At the start of an analysis, it's common to select all columns, or in SQL, use a select star query. That's okay if you're just beginning, but once you know your goal or at least how to reach it, go back and remove any select-all statements. Most analytics ultimately utilize only a few columns. Be particular!

It's tempting to include all the data because you never know when you'll need something, right? At first, you might select many columns, but after you're done, pause to reflect – how much mental burden do you want to place on your future self? This is the communication effect of the DFM – be clear with your choices, and your future users will thank you, as seen in Table 3-5.

27

CHAPTER 3   DATA FLOW MAP DEEP DIVE

Table 3-5. *Cut Potential*

@	Choices Matter	Pick only the variables that change the world
O	Get	Coffee Data
V	Date	When did the purchase happen?
V	Amount	How much was it worth?

› Slice

Where a Cut removes one or more columns, a Slice removes one or more rows of data. Slicing cares about what the data is – this value or that, a particular date, or a range of numbers. In Figure 3-5, I'm focusing on rows 1 and 3, dropping row 2.

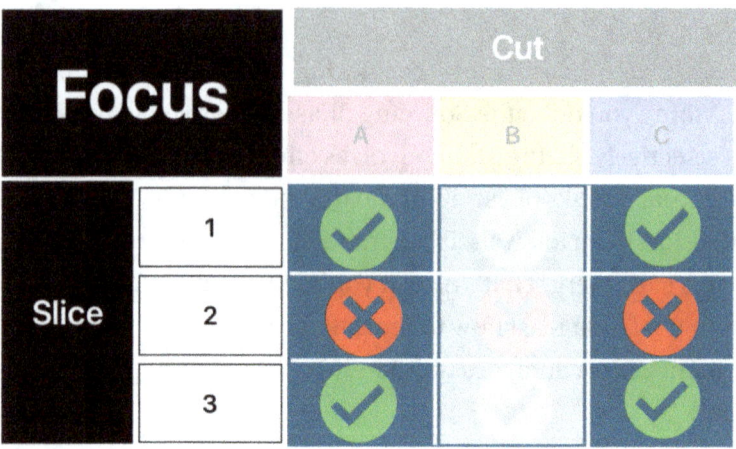

Figure 3-5. *Focus Mode Slice*

As humans, we're limited in how many rows we can handle mentally. And honestly, those limits are quite small. Give our brains a break – cut down on the data as much as possible, shown in Table 3-6.

28

Table 3-6. *Slicing Options*

@	Convention
>	Row Limit
>	Conditional Value
>	Nulls / Blanks
>	Lists
>	Sampling

Let's explore each of these in more detail.

Row Limit

The easiest and most direct way to prevent overloading yourself with data is to set a firm limit at the start of the code. By reducing the processed volume, it becomes simpler to clearly define what data is available. Once the process is verified, remove these temporary limits. A "LIMIT 10" isn't arbitrary (more on that below); it encourages thought and shows whether the analysis is on track. Importantly, many analytic tools will run much faster when they don't overload the computer, making it easier for you to be creative.

Selecting all the columns to check is the most effective way to get a quick result. This taste test can show you if you have what you need to move forward or if you need to build multiple stages with Fixes or Mixes that strengthen the foundation. This might look like Table 3-7.

Table 3-7. Select First Ten Rows

@	Short List
0	Massive Dataset
>	First Ten Rows

Depending on the system, you might incorporate some wildcard patterns to select a range of values that match more general conditions. See Table 3-8.

Table 3-8. Wildcarding

@	Patterns in the sand
0	Massive Dataset
>	Color starts with "Red"
>	Name is like "Tom"

Conditional Value

Sometimes, you care about volume – how many of this, how few of the other. These are exercised against the data's metric, blocked out in Table 3-9. This is most commonly expressed in some variation of "WHERE ..." or in table filters. For example, my SQL query might read:

```
SELECT *
FROM source
WHERE sales_amount = 1.00
AND weight = 5
AND time = "May"
```

CHAPTER 3 DATA FLOW MAP DEEP DIVE

Table 3-9. *Values Matter*

@	Specifics
0	Massive Dataset
>	Weight is Five Pounds
>	Cost is One Dollar
>	Time is Last Month

Lists

Similar to conditional values, lists identify information, usually in a binary on/off format. You can create lists that include multiple matches, making it easy to capture several useful conditions. See Table 3-10.

Table 3-10. *Multiple Values*

@	Specifics
0	Massive Dataset
>	Color is Green, Yellow, Red
>	Size is Small or Large

Nulls/Blanks

Nulls can be a whole chapter on their own. A null doesn't mean zero; it means, "I ... don't know!" Nulls often indicate gaps in data or integrity issues. Where left joins are used, they can show that related rows are missing. See Table 3-11.

CHAPTER 3 DATA FLOW MAP DEEP DIVE

Table 3-11. *Not Sure?*

@	Specifics
0	Massive Dataset
>	Is Null – No Known Value
>	Is not Null – Has a Value

Sampling

When you're first testing and putting together your data pipeline, setting up a small, randomized sample of the data (perhaps 100 rows at most, or even ten) is the best way to exercise the code and receive fast, immediate feedback from the system. It provides you with immediate clarity on the value of your data before you begin running a much larger analysis. If you do find issues in the sample, you can then go back and Fix the columns that need updates.

In other cases, the overall volume is so large or your actual time limits are so tight that you must use a randomized selection. While computer capacity is increasing, so are our desires to handle larger datasets and generate deeper insights. See Table 3-12.

Table 3-12. *Random Selection*

@	Specifics
0	Massive Dataset
>	Random Row Selection

CHAPTER 3   DATA FLOW MAP DEEP DIVE

\ Sort

Sorting data is both a column and a value choice; you must pick a column to sort and values within that column, either from top to bottom or bottom to top. In Figure 3-6, I'm sorting Column A from high values to low (commonly referred to as descending) and Column C in the ascending order. Column B is left untouched.

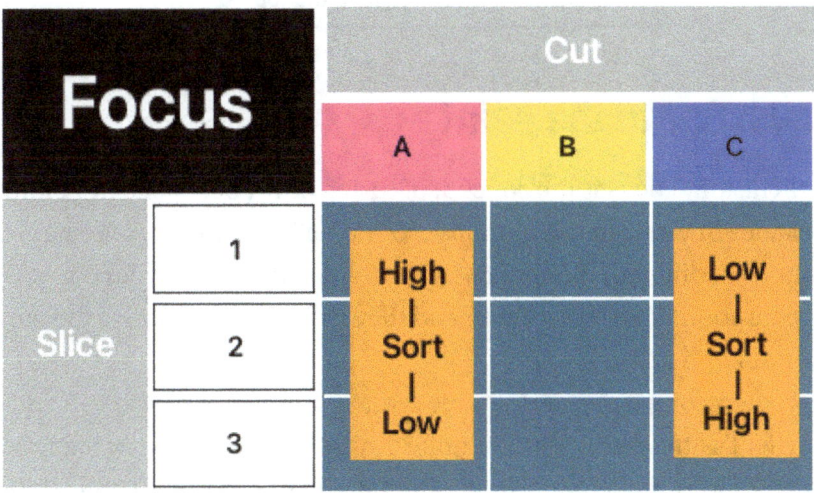

Figure 3-6. Sorting Is a Choice of Values and Columns

Sorting feels like the least complex feature of analytics, and yet it's often the most important because it means you focus on the most significant things first. How you rank those things may be largest to smallest or the other way round, but you want to be able to quickly re-order data, often on multiple criteria at the same time. Get familiar with your analytic tool of choice's approach to sorting to help focus on the key issues quickly, as sketched out in Table 3-13.

Table 3-13. Sorting by Price and Then Size

@	Specifics
O	Massive Dataset
>	Select closest locations first
>	Then select price, low to high
>	Finally limit the results to the top 10

Focus: Get to the point

In conclusion, Focus is the concept of caring only about specific columns and values in your dataset. Suppose your data is extensive, covering hundreds of different dimensions. That's great at first, but your client will only care about a few key values, and they will always prioritize the most important one first.

Modern computing and storage seem unlimited, but both you and your client's attention spans are quite limited. Streamline your analysis to quickly present the most important data, so minimal time and effort are spent understanding the key information needed for the next step – Build mode.

You have a Source of data, and you have Focused on the most significant bits. Now, let's Build something valuable from these selected materials.

Build

Sourcing and Focusing prepare you for the big show: creating something new from specific raw materials. Building new, useful analytics is about three key actions:

- **Box**: Finding categories and grouping data accordingly
- **Size**: Measuring the impact of those boxes
- **Ship**: Creating a new result

Box allows us to categorize the world, emphasizing the essential. Analytics truly reaches its potential when you begin generating new, actionable insights that you and your business partners can use to view the world differently and then change it.

After constructing a new Framework around your data, understanding the data's size relative to everything else reveals the significance of its impact.

Wondering if this brand outperforms the other? Measure sales. Is this technique cheaper than another? Measure its effects. Do you want someone to change their behavior? Give them one clear and important measure that they can understand and act on.

Finally, after you've completed the entire process of Sourcing, Focusing and Building summaries, it's time to deliver. None of this matters until you can present that creative, laser-focused analysis to the world and make a difference. By using these actions, you will create clarity around your analytics, making it easier to gain adoption.

] Box

Boxing data revolves around commonalities – what unites separate groups. These could encompass complex factors, such as diversity, or a simple choice of car color, brand, and model.

The Box action functions as a way of identification; we classify objects with labels. Part of the Sourcing and Fixing process involves preparing you to use those artificial labels effectively. It encourages us to recognize the dividing lines between this and that.

To illustrate this process, I'll use simple graphics – on the left are the raw, random data represented by circles, squares, and triangles, and on the right is the outcome of a box, as shown in Figure 3-7. In this case, I've bucketed the data by shape into similar categories.

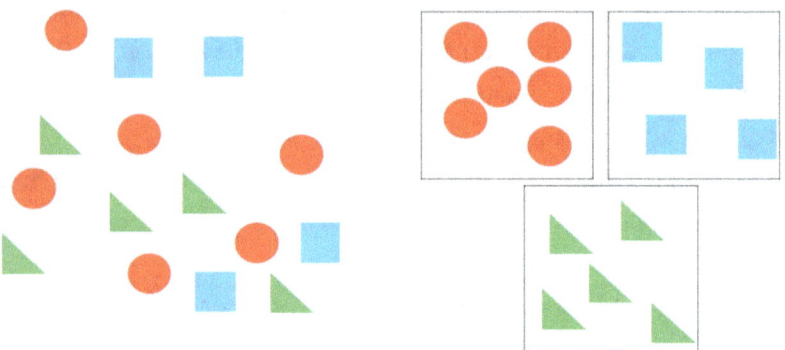

Figure 3-7. Little Boxes

Size

The natural complement to a Box is its Size. Whatever lines define the Box's boundaries, consider the measurement of the items contained within.

This could be dimensional (height x width x length), weight, or count. Wherever there's a Box, there will always be a count. And, conversely, almost always, where there's a count, there's a box. See Figure 3-8.

CHAPTER 3   DATA FLOW MAP DEEP DIVE

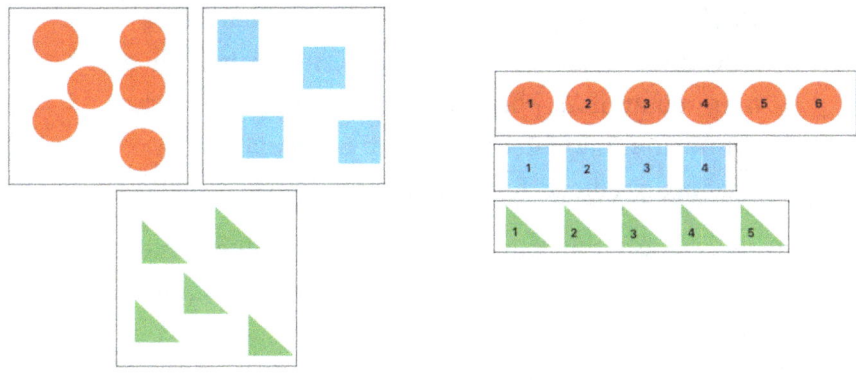

Figure 3-8. Counting Shapes

Size whether large or small, can tells us two things:

1. Does it matter? Once achieved, does it shake the world to its core, or does a collective yawn suggest it was a reasonable effort but perhaps other things could have been more worthwhile for the world?

2. How much effort will this require? Is it simple and do we have the resources (people, time, money)?

When taking the measure of something, we tend towards these mistakes:

- **Too Many Rows**: An ocean of numbers spread across a sea of indifferent data. There's an opportunity for the analyst to cut through the noise by summarizing the metrics. The reader needs to know where to look and the analyst must take the time to identify what might be crucial.

- **Having Too Many Metrics**: Often seen when feeling insecure – the analyst overcompensates by including as many columns as possible, confusing the reader who cannot identify which metrics, in particular, contain the crucial ingredients for success.

Sizing the shape of the data – whether it's summing naturally additive numbers, averaging trending indicators, or counting unique items in categories – is the "how much" of analysis. Every analysis requires a metric, even if that metric serves as reassurance for the analyst that things are functioning as expected. Numbers allow us to say no.

Let's explore a few different ways an analysis can measure impact.

Statistically Speaking

Some numbers can never be added; they are thoughtfully examined over time. Temperature, speed, and altitude all matter, but you can't add up the fact that I was driving 50 mph 10 minutes ago, and now I'm driving 75 mph – unless you're an officer of the law, in which case I had no idea, and here's my insurance.

Stats to the rescue – and they don't have to be deep data scientist stats either. The basics can take you a long, long way:

- **Min**: The least value
- **Max**: The highest value
- **Med**: The median value – string up the numbers sequentially on a line and pick the middle one by count of data points
- **Avg**: Average (or mean) – this number may be a fraction, but it's the middle mathematical value

Potential Metrics

Basic Math is the simplest metric and most easily understood by your audience.

- **Count Rows**: You can go a long way just knowing how often an event occurs.
- **Count *Distinct* Values**: Selling products? How many of each type of product.
- **Sum Up Anything That's a Measure**: Go crazy; you never know when it'll come in handy.

Numbers As Labels

Some numbers are recognized as facts. They are all labels or names for something or someone, such as social security numbers, barcodes, and group numbers.

These identifiers are always useful for a Box and are typically better handled as strings rather than numbers.

= Ship

Like the period at the end of a sentence, this is the briefest moment, yet just as important. Leave your readers wanting more but clear about where they stand in the universe. Even if this is just one step in a long journey, be upfront and specific about what you're creating. Later, when you come back to this cold, the final result may be where you start to trace how some result came to be.

Thoroughly documenting the final destination greatly enhances clarity in an analysis, making it much easier and more accurate to communicate the process to another analyst or reader.

Where an analysis stage or entire process concludes is often similar to where it is recorded in a Get in the Source stage. The following are the most common ways to complete an analytic.

Save to File

While the CSV file is the most common format, you can place your data in various locations and formats. But seriously, folks, why would you use anything else?

Comma separations are fine but are prone to issues with text fields that include commas. Therefore, these must be wrapped in quotes, resulting in sometimes bizarre bits of code. On the other hand, you may need to save this to an Excel file for a future analyst or perhaps a Parquet file for easy import into a data mining workflow.

Memory

If the data will be used in a future analytical stage as a Source, simply keeping it in memory is ideal. The chosen analytical tool will often manage file handling, memory, and Sourcing more clearly and easily.

In a database, Common Table Expressions are SQL's way of organizing your code more clearly and efficiently. In this case, the outcome is almost always the same as the handle. Alternatively, you might be storing data in a data frame that supports a larger goal.

The result is your handle for the future, which is then used in the final query. Each part can focus on one step of the process – maybe step one is gathering and cleaning, step two is boxing and sizing, and you sprinkle cuts throughout the process.

See

This is the stunning Technicolor result of all your hard work.

Charts and visualizations are the pinnacle of analysis, and rightly so. They condense thousands or millions of data points into a single, easy-to-understand grid of insights, empowering you to make bold moves. Please do it!

However, that doesn't mean the visual is easily understood by your business partners and clients. Whenever possible, use the Data Flow Map technique on your charts to help your viewers grasp what's happening behind the technical details.

Build: Shipping a Conclusion

Initially, you have raw materials. You shape them, combine them with other sources, and make some tough choices, focusing on the key point of the analysis. Then you build towers of categories and rate and rank them by size. Your audience will be impressed, and your portfolio will grow with one more impressive analysis.

If your analytic process is complex, it's important to use a straightforward tagging system to handle all its stages. As the analysis develops into something more useful, use simple tags and be prepared to update or remove them as necessary.

In this final section, I'll examine what makes a good tag before wrapping everything up.

Tag

As you work through your analysis by building chunks of sources, choices, and structures, you will need a way to organize them into clear stages. Stage tags will act as labels for a complex analysis. They serve as memory cues your brain uses to track what's important, what your thought process was, and to ignore details that aren't relevant right now.

Tagging is a form of nickname – they're intentionally short, yet they help your mental concepts stand out.

What's in a Tag?

Everything and nearly nothing. It should be brief, perhaps a word or two, three at most.

This is a reasonable Tag:

@ Collect Weather Stats

This is a **bad** one:

@ Go to the weather site and download the weather data, saving it to...

That's doing the job of the analysis. Here's another **bad** one:

@ Get

While this is technically an Action, it's not specific enough to understand the purpose.

A tag needs to be something you can carry forward, being a placeholder for a broader, well-defined set of steps.

Finally, don't take it too seriously. This will not grow up to carry on your legacy; it is simply a block of thought in your head that must be unique to you. Be kind to your data pipeline.

Examples

See Table 3-14 for a list of good tags:

Table 3-14. Sample Tags

Examples
Collect Weather Stats
Clean up stat data
Update Weather History
Create Today's Snapshot
Process Results Reports

Keyword Conflicts

Every analytic language, SQL being no exception, has a laundry list of keywords that you should not use as table, field, or expression names. These include the obvious SELECT, WHERE, and FROM, as well as surprising edge cases around adjectives that might be potential candidates for a tag.

While keywords can act as tags, they often cause confusion between the technical process and the Data Flow Map. Your coding mindset benefits from concise, impactful words that specify source, action, and control. It's tempting to label a section of code as GROUP, but then you need to translate back and forth to keep everything clear, and that's exactly what the Data Flow Map is meant to help with.

Get creative, get silly, go outside, and make them memorable. If I ask you the color of the red car and you say, "Red," you've correctly answered the question with the facts. If the answer is "Bloody Lipstick Red," that's an entirely different kettle of fish, and you can bet I will remember this for a very long time.

CHAPTER 3   DATA FLOW MAP DEEP DIVE

The tag should be the first thing you choose and the last thing you update. Using linking language, you can outline an entire multi-stage analysis using tags alone.

Summary

Remember the modes: Source, Focus, and Build. Each mode has three actions. Complex analytics can be divided into multiple stages, each marked by a simple tag. See Figure 3-9.

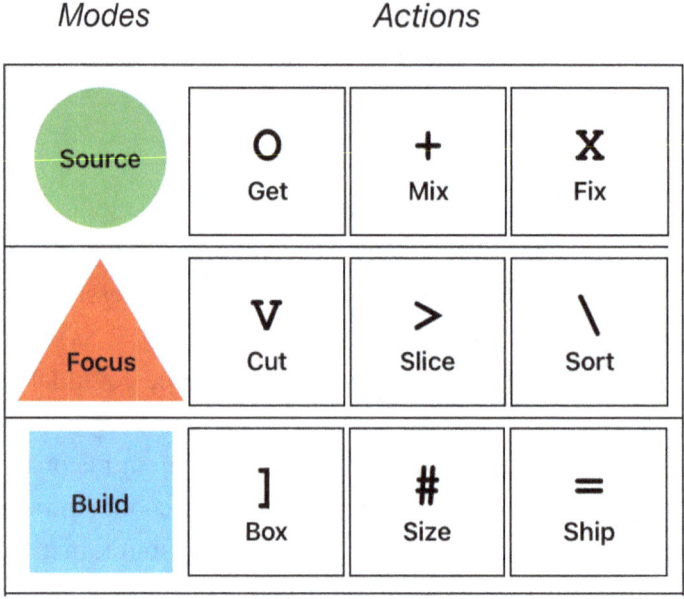

Figure 3-9. The Data Flow Map Framework

If you are faced with a new analysis and don't know where to begin, try to identify the key stages it goes through. Grab some tags and apply them to each stage, each of which will start with Sourcing, narrow the scope with Focusing, and Building something novel and new.

Are you inheriting someone else's code, spreadsheets, or SQL queries? Break them into logical chunks, identify where they gather data, highlight their decisions, and use the Data Flow Map to be more creative, clarify the analytic process, and effectively communicate the results to your team members and clients.

The book's next section will explore several example datasets using various tools. Hopefully, some will be familiar, and others will inspire you to broaden your skills and abilities with analytic tools.

The following content is broken into the following categories:

- **Files**: Spreadsheets, CSV, and binary
- **Databases**: Traditional RDBM and newer document
- **Code**: Hitting the path with Python or other languages
- **APIs**: Command line tools
- **Platforms**: All-inclusive beach hotels of analytic tools
- **Pipelines**: Visual tools for data engineering
- **Analog**: Pen and paper applied to real life

Sample data will come from various simple sources, all of which are included in the Appendix of this book and on GitHub.com.

CHAPTER 4

Examples – Files

Files

In this chapter, I will guide you through the complete process of Sourcing, Focusing, and Building analytical results to meet your objectives, all within a single spreadsheet.

In analytics, file-based sources are typically the most straightforward. You open a file on your computer with a specific tool, frequently a spreadsheet application, manipulate the data, generate results, and save them back to the file.

The traditional spreadsheet approach is familiar to billions of computer users worldwide, regardless of their titles. While this chapter will use Microsoft Excel, the most popular spreadsheet program, the principles apply to any similar software capable of handling files.

Comma-separated Value (CSV) files are the most common format in the world, followed by the Excel format itself. Excel automatically opens them on your computer, as will nearly every other tool, platform, or plain text editor.

Ultimately, this aims to explore the Data Flow Map Framework rather than serve as an authoritative guide for managing Excel data. Your version may differ slightly, and many online resources are available to assist you in getting started.

CHAPTER 4   EXAMPLES – FILES

I'll examine sample weather observation data and include a cloud cover type for context. The data is stored in a spreadsheet called "Weather_Log.xlsx," which contains two worksheets: "Weather History" and "Cloud Cover." The goal is to categorize the data by cloud cover type, number of days, and average temperature.

Source

The initial data sourcing in a spreadsheet is the easiest part; ironically, joining data is often the hardest.

o Get

Using my File Explorer, open the file, as in Figure 4-1.

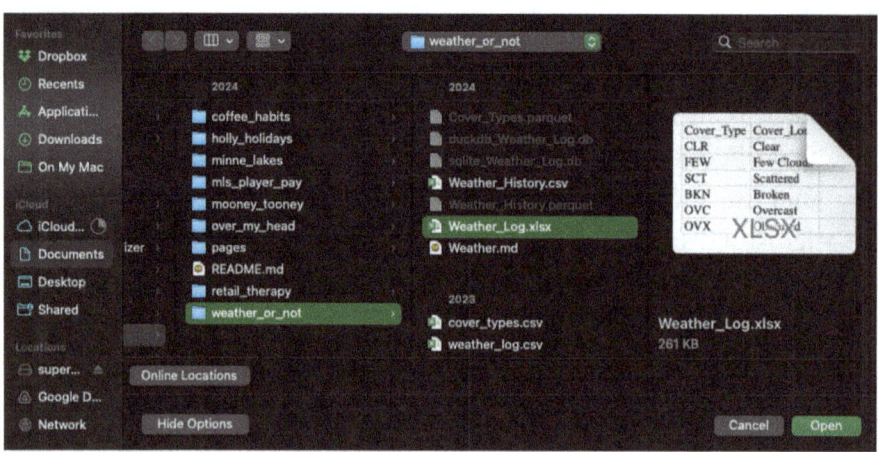

Figure 4-1. *Get a File*

CHAPTER 4 EXAMPLES – FILES

This should result in the worksheet "Weather History"; Figure 4-2.

	A	B	C	D	E	F	G	H	I
1	Type	Station	Observed_D	Temp_Celsi	DewPoint_C	Wind_Speed	Wind_Direct	Cover	Visibility
2	METAR	KMSP	2022-12-27T	-15.6	-20	5	220	BKN	10
3	METAR	KMSP	2022-12-27T	-15.6	-20	5	220	BKN	10
4	METAR	KMSP	2022-12-27T	-16.1	-21.1	5	200	SCT	10
5	METAR	KMSP	2022-12-27T	-16.1	-20.6	0	0	BKN	10
6	METAR	KMSP	2022-12-27T	-16.1	-20.6	6	180	BKN	10
7	METAR	KMSP	2022-12-27T	-16.1	-20.6	5	170	BKN	10
8	METAR	KMSP	2022-12-27T	-16.1	-20.6	7	150	OVC	10
9	METAR	KMSP	2022-12-27T	-15	-20	7	160	OVC	10
10	METAR	KMSP	2022-12-27T	-13.9	-19.4	7	130	BKN	10

Figure 4-2. Weather History

+ Mix

In the same workbook, there's another Worksheet called "Cover_Types"; Figure 4-3.

	A	B	C	D
1	Cover_Type	Cover_Long	Max_Coverage_pct	
2	CLR	Clear	0	
3	FEW	Few Clouds	0.25	
4	SCT	Scattered	0.5	
5	BKN	Broken	0.88	
6	OVC	Overcast	1	
7	OVX	Obscured		

Figure 4-3. Cover Types

CHAPTER 4   EXAMPLES – FILES

I'll use the "Cover_Long" column to help decode the "Cover" column in the primary data; highlighted in Column H in Figure 4-4.

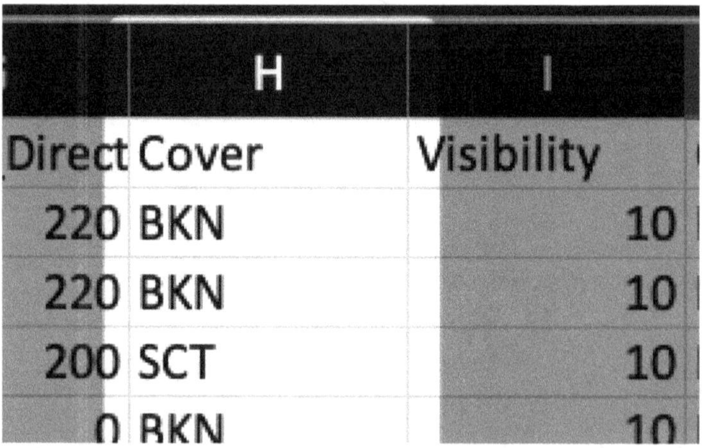

Direct Cover	Visibility
220 BKN	10
220 BKN	10
200 SCT	10
0 BKN	10

Figure 4-4. Highlight Cover Column

Joining data in Excel using formulas is usually done with the VLOOKUP function. Note: Excel includes the Power Query tool, a more powerful way to manage data. While this guide won't cover that tool, it's worth exploring as an alternative to VLOOKUP and other data processing tools.

A VLOOKUP formula includes the following elements:

1. **Lookup Value**: What are you decoding?

2. **Table Array**: Where are you getting the lookup values from? In Excel, this must be a fixed reference (using dollar signs) to prevent errors.

3. **Col Index Num**: Which column of the lookup table are you picking?

4. **Range Lookup**: Always False for an exact match.

See Figure 4-5 for the formula method and Figure 4-6 for the result.

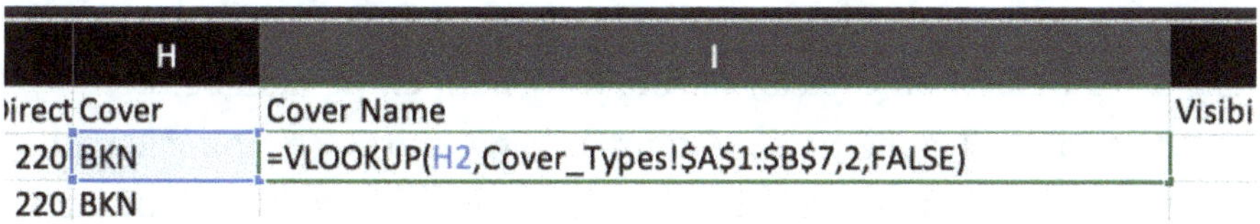

H	I	
)irect Cover	Cover Name	Visibi
220 BKN	=VLOOKUP(H2,Cover_Types!$A$1:$B$7,2,FALSE)	
220 BKN		

Figure 4-5. Complete Formula

Cover	Cover Name	Coverage Perc
BKN	Broken	0.88
OVC	Overcast	1
OVC	Overcast	1
FEW	Few Clouds	0.25
OVC	Overcast	1
BKN	Broken	0.88
FEW	Few Clouds	0.25
CLR	Clear	0
FEW	Few Clouds	0.25
BKN	Broken	0.88

Figure 4-6. Result

x Fix

Now that I've decoded the Cover code, let's fix the data in the Date column. Currently, it's a string with the year, month, day, the letter "T," the time, and then "Z," which indicates Coordinated Universal Time (UTC). I want to separate the date and time from this and adjust the time for Central Time, United States, so I can focus only on the entries for noon.

CHAPTER 4   EXAMPLES – FILES

This is an area where Excel shines – it's reasonably easy to use the built-in formulas, and you can see the work happening right before your eyes. First, I'll split the date stamp into two parts. I'm going to add two blank columns following the date stamp and then start with this in Figure 4-7.

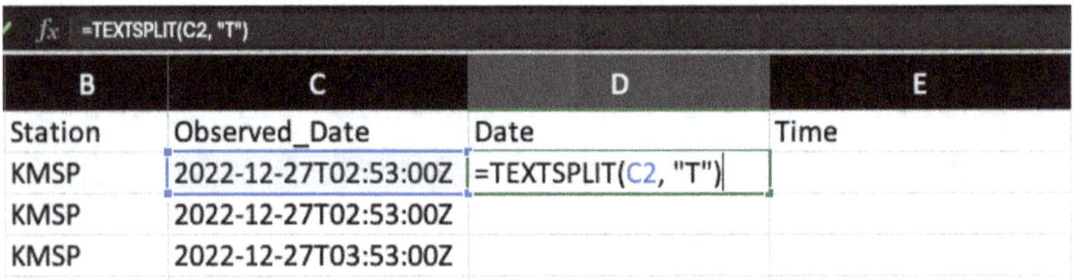

B	C	D	E
Station	Observed_Date	Date	Time
KMSP	2022-12-27T02:53:00Z	=TEXTSPLIT(C2, "T")	
KMSP	2022-12-27T02:53:00Z		
KMSP	2022-12-27T03:53:00Z		

Figure 4-7. Split the Date

This results in Figure 4-8.

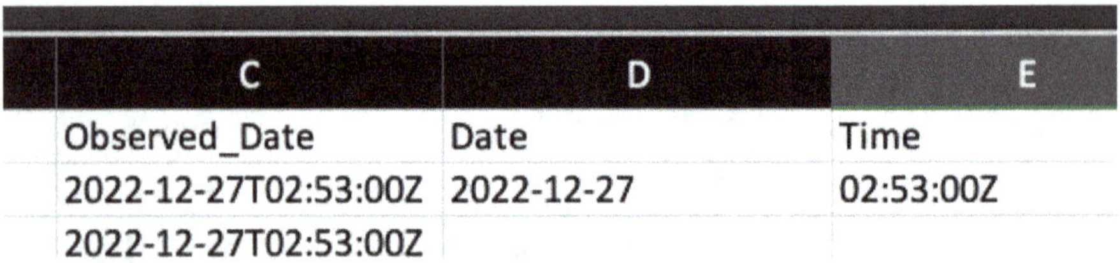

C	D	E
Observed_Date	Date	Time
2022-12-27T02:53:00Z	2022-12-27	02:53:00Z
2022-12-27T02:53:00Z		

Figure 4-8. Banana Split Result

I still need to remove the "Z" from the end of the second value, so I'll add a third column and remove it in Figure 4-9.

CHAPTER 4   EXAMPLES – FILES

E	F	G
Time	Time Clean	Temp_C
02:53:00Z	=REPLACE(E2, 9, 1, "")	-.
02:53:00Z	02:53:00	-.
03:53:00Z	03:53:00	-.
04:53:00Z	04:53:00	-.

Figure 4-9. Remove the Unnecessary Z

Since I have to adjust for the time zone, I'd like to convert the remaining value into the Time value that Excel understands in Figure 4-10.

E	F	G	
Time Half	Time No Z	Time	Ter
02:53:00Z	02:53:00	=TIMEVALUE(F2)	
02:53:00Z	02:53:00		

Figure 4-10. Time Time

This is shown in Figure 4-11.

C	D	E	F	G
Observed_Date	Date	Time Half	Time No Z	Time
2022-12-27T02:53:00Z	2022-12-27	02:53:00Z	02:53:00	2:53:00 AM
2022-12-27T02:53:00Z	2022-12-27	02:53:00Z	02:53:00	2:53:00 AM
2022-12-27T03:53:00Z	2022-12-27	03:53:00Z	03:53:00	3:53:00 AM
2022-12-27T04:53:00Z	2022-12-27	04:53:00Z	04:53:00	4:53:00 AM
2022-12-27T05:53:00Z	2022-12-27	05:53:00Z	05:53:00	5:53:00 AM
2022-12-27T06:53:00Z	2022-12-27	06:53:00Z	06:53:00	6:53:00 AM
2022-12-27T07:53:00Z	2022-12-27	07:53:00Z	07:53:00	7:53:00 AM
2022-12-27T08:53:00Z	2022-12-27	08:53:00Z	08:53:00	8:53:00 AM

Figure 4-11. Real-Time

CHAPTER 4 EXAMPLES – FILES

For this analysis, I will ignore that adjusting back six hours will create some funny results, but this should work for the demonstration. If you would use this on actual data, consider incorporating the date into the date math. Date math is often tricky and confusing. Test early and often.

For now, let's subtract six hours to get it roughly into my time zone in Figure 4-12.

F	G	H	
Time No Z	Time	Time Local	Te
02:53:00	2:53:00 AM	=G2-time(6,0,0	
02:53:00	2:53:00 AM	TIME(hour, minute, second)	
03:53:00	3:53:00 AM		

Figure 4-12. *Simple Time Math*

Okay, that didn't work as expected; see Figure 4-13.

F	G	H	T
Time No Z	Time	Time Local	
02:53:00	2:53:00 A	##########################	
02:53:00	2:53:00 AM	##########################	
03:53:00	3:53:00 AM	##########################	
04:53:00	4:53:00 AM	##########################	
05:53:00	5:53:00 AM	##########################	
06:53:00	6:53:00 AM	12:53:00 AM	
07:53:00	7:53:00 AM	1:53:00 AM	
08:53:00	8:53:00 AM	2:53:00 AM	
09:53:00	9:53:00 AM	3:53:00 AM	
10:53:00	10:53:00 AM	4:53:00 AM	

Figure 4-13. *Wait a Minute*

It looks like anything under six hours is resulting in an error. It's creating a negative time, which Excel doesn't support. To fix this, let's use the MOD function to wrap the clock around 24 hours; Figure 4-14.

CHAPTER 4　EXAMPLES – FILES

F	G	H
Time No Z	Time	Time Local
02:53:00	2:53:00 AM	=MOD(G2-TIME(6,0,0),1)
02:53:00	2:53:00 AM	##########################
03:53:00	3:53:00 AM	##########################
04:53:00	4:53:00 AM	##########################
05:53:00	5:53:00 AM	##########################
06:53:00	6:53:00 AM	12:53:00 AM
07:53:00	7:53:00 AM	1:53:00 AM
08:53:00	8:53:00 AM	2:53:00 AM

Figure 4-14. *Add the MOD*

In this case, I'm stacking the formulas, which is fine, but remember, it's easy to get lost in the parentheses, commas, and options. It's better to keep breaking it out into sibling columns. Nonetheless, the result is good; Figure 4-15.

C	D	E	F	G	H
Observed_Date	Date	Time Half	Time No Z	Time	Time Local
2022-12-27T02:53:00Z	2022-12-27	02:53:00Z	02:53:00	2:53:00 AM	8:53:00 PM
2022-12-27T02:53:00Z	2022-12-27	02:53:00Z	02:53:00	2:53:00 AM	8:53:00 PM
2022-12-27T03:53:00Z	2022-12-27	03:53:00Z	03:53:00	3:53:00 AM	9:53:00 PM
2022-12-27T04:53:00Z	2022-12-27	04:53:00Z	04:53:00	4:53:00 AM	10:53:00 PM
2022-12-27T05:53:00Z	2022-12-27	05:53:00Z	05:53:00	5:53:00 AM	11:53:00 PM
2022-12-27T06:53:00Z	2022-12-27	06:53:00Z	06:53:00	6:53:00 AM	12:53:00 AM
2022-12-27T07:53:00Z	2022-12-27	07:53:00Z	07:53:00	7:53:00 AM	1:53:00 AM
2022-12-27T08:53:00Z	2022-12-27	08:53:00Z	08:53:00	8:53:00 AM	2:53:00 AM
2022-12-27T09:53:00Z	2022-12-27	09:53:00Z	09:53:00	9:53:00 AM	3:53:00 AM
2022-12-27T10:53:00Z	2022-12-27	10:53:00Z	10:53:00	10:53:00 AM	4:53:00 AM

Figure 4-15. *Final Date and Time Stamps*

Since I'm observing data in the United States, I'm more comfortable looking at the temperatures in Fahrenheit rather than Celsius, so I will apply the classic conversion formula in a new column. In addition, I'd like to round out the numbers to whole digits. See Figure 4-16.

CHAPTER 4   EXAMPLES – FILES

	I	N	
	Temp_Cels	Temp_Far	Cover Name
0 PM	-15.6	=ROUND((I2 * 9/5) + 32,0)	
0 PM	-15.6	4	Broken

Figure 4-16. *Fahrenheit Conversion*

This should give the result in Figure 4-17.

H	I	N	
	Temp_Celsius	Temp_Farenheit	Cover Name
8:53:00 PM	-15.6	4	Broken
8:53:00 PM	-15.6	4	Broken
9:53:00 PM	-16.1	3	Scattered
10:53:00 PM	-16.1	3	Broken
11:53:00 PM	-16.1	3	Broken
12:53:00 AM	-16.1	3	Broken
1:53:00 AM	-16.1	3	Overcast
2:53:00 AM	-15	5	Overcast
3:53:00 AM	-13.9	7	Broken
4:53:00 AM	-13.3	8	Overcast
5:53:00 AM	-12.2	10	Overcast
6:53:00 AM	-11.7	11	Scattered
7:53:00 AM	-11.7	11	Broken
8:53:00 AM	-9.4	15	Broken
9:53:00 AM	-8.3	17	Broken
9:53:00 AM	-8.3	17	Broken

Figure 4-17. *Fixed Temps*

All this prep work in Excel is okay because, in the "Focus" section, we will ignore the in-between columns. However, if you have a considerable dataset, this can become cumbersome.

Now I've got some valuable data that I can start building with. However, I've got a lot of data that doesn't matter. Let's hone in on the important stuff.

CHAPTER 4   EXAMPLES – FILES

Focus

I have the basics of the data and have joined two datasets. Unfortunately, there is a lot more data than I need. I can do something about that.

v Cut

This dataset now has 17 columns. I only want to focus on:

- Date
- Time
- Cover
- Temperature

If there's not a lot of data, and there's not, let's hide the irrelevant columns by selecting them and then right-clicking the columns. See Figure 4-18.

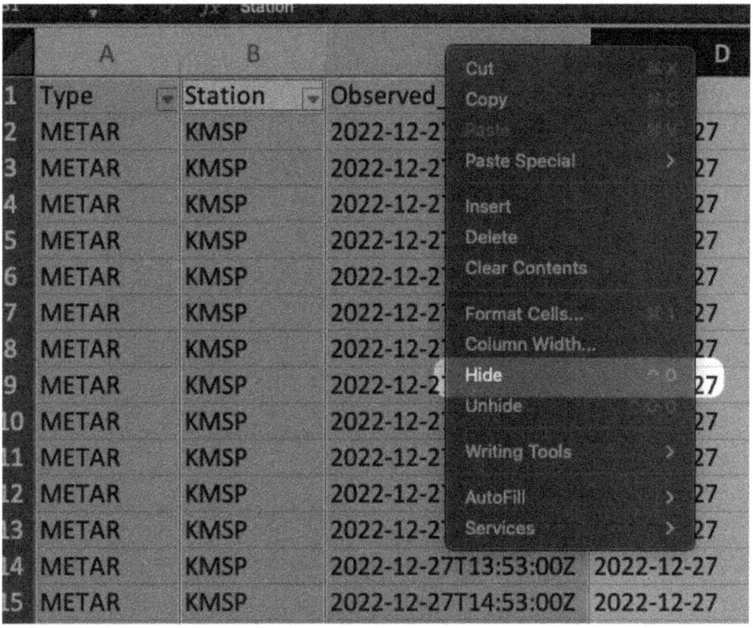

Figure 4-18. Hiding the Columns

This should give the following much cleaner point of view in Figure 4-19.

Date	Time Local	Temp_Cels	Cover Name
2022-12-27	8:53:00 PM	-15.6	Broken
2022-12-27	8:53:00 PM	-15.6	Broken
2022-12-27	9:53:00 PM	-16.1	Scattered
2022-12-27	10:53:00 PM	-16.1	Broken
2022-12-27	11:53:00 PM	-16.1	Broken
2022-12-27	12:53:00 AM	-16.1	Broken
2022-12-27	1:53:00 AM	-16.1	Overcast
2022-12-27	2:53:00 AM	-15	Overcast
2022-12-27	3:53:00 AM	-13.9	Broken
2022-12-27	4:53:00 AM	-13.3	Overcast
2022-12-27	5:53:00 AM	-12.2	Overcast
2022-12-27	6:53:00 AM	-11.7	Scattered

Figure 4-19. *Just the Columns I Need*

> Slice

Now that I have removed unnecessary columns, I can reduce the rows to what matters.

In Excel, I can select the data I want from the filter button on the table headers. Figure 4-20 shows how.

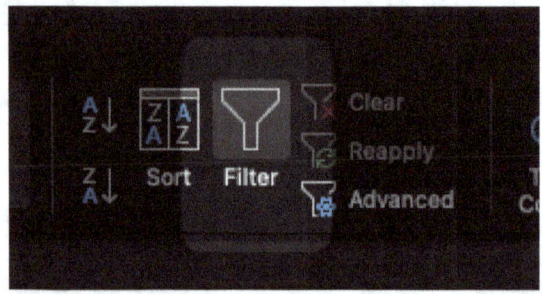

Figure 4-20. *Apply Filters*

Now that I have applied filter handles, I can hit the drop-down for any column. In this case, I only want to pick rows that are at 11:53 AM; Figure 4-21.

CHAPTER 4   EXAMPLES – FILES

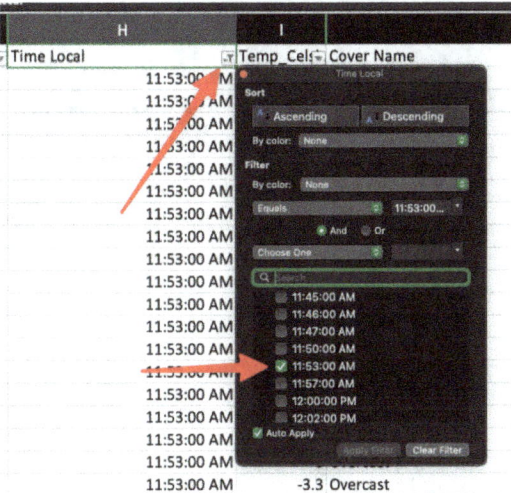

Figure 4-21. *Picking Just One Time*

The reason I want this is a little obscure. Usually, the regular weather posts just before the hour mark, with multiple alerts spread around the rest of the hour. Since I'm looking for regular patterns, I'll only focus on the regular updates.

The weather station is recording the weather at the 53rd minute of the hour. See the result in Figure 4-22.

Date	Time Local	Temp_Farenheit	Cover Name
2022-12-27	11:53:00 AM	20	Broken
2022-12-28	11:53:00 AM	35	Overcast
2022-12-29	11:53:00 AM	35	Overcast
2022-12-30	11:53:00 AM	21	Few Clouds
2022-12-31	11:53:00 AM	30	Overcast
2023-01-05	11:53:00 AM	26	Broken
2023-01-06	11:53:00 AM	13	Few Clouds
2023-01-07	11:53:00 AM	10	Clear
2023-01-08	11:53:00 AM	18	Few Clouds
2023-01-09	11:53:00 AM	24	Broken
2023-01-10	11:53:00 AM	25	Overcast
2023-01-11	11:53:00 AM	31	Overcast

Figure 4-22. *On the 53's*

59

CHAPTER 4   EXAMPLES – FILES

Now that I've got just the rows I need, I can drop the "H Time Local" column; Figure 4-23.

Date	Temp_Farenheit	Cover Name
2022-12-27	20	Broken
2022-12-28	35	Overcast
2022-12-29	35	Overcast
2022-12-30	21	Few Clouds
2022-12-31	30	Overcast
2023-01-05	26	Broken
2023-01-06	13	Few Clouds

Figure 4-23. *Cleaner Data*

The columns are tight, and the rows are right. Does the data feel correct? I'll use sorting to help me determine that.

\ Sort

The data is organized chronologically, so I'm not concerned about that. However, conducting the top/bottom value testing through sorting is useful. This is a tiny dataset, and I'm not putting too much pressure on the system – if it were millions of rows, I'd reconsider the platform and value testing. I would begin by selecting random rows to reduce the dataset to a manageable size that quickly provides results.

Sorting is typically one of the final touches just before Shipping the data – make sure the client sees the most critical information first. For this reason, I'm developing a mental model for the range of data.

Click the down arrow at the top of the column to sort it in both directions and observe the results. Repeat this process for all columns. See Figure 4-24.

CHAPTER 4   EXAMPLES – FILES

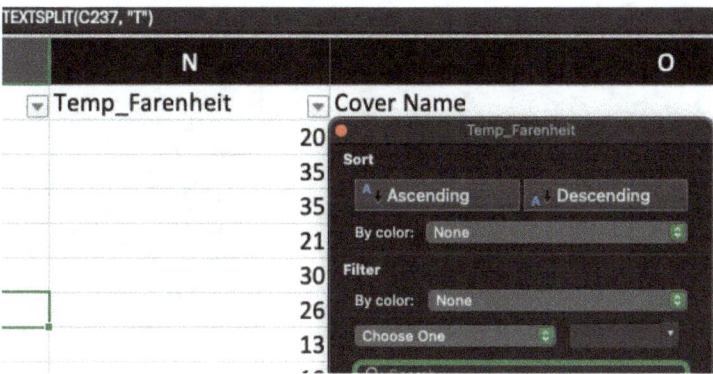

Figure 4-24. The First Taste Test

This results in Figures 4-25 and 4-26.

Figure 4-25. Temp Low to High

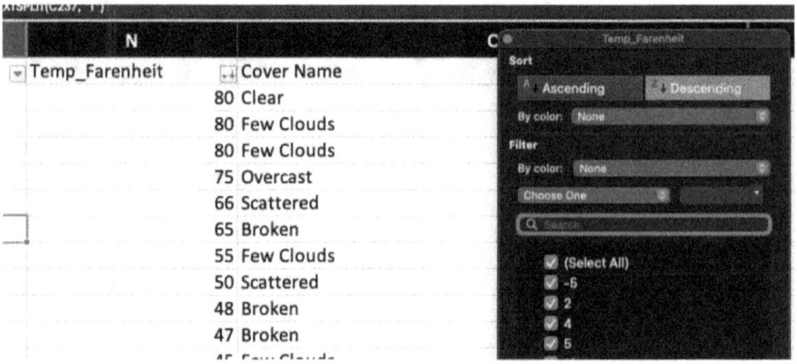

CHAPTER 4   EXAMPLES – FILES

Figure 4-26. Temp High to Low

That's quite a range, stretching from -5 degrees to 80 degrees Fahrenheit. It essentially goes from the dead of winter to a very warm spring. You can almost see the coats being tossed back into the closet.

I will repeat this process with the Cover Name and Date. These quick tests almost immediately indicate the range of data I am dealing with and confirm that I am covering from the end of December to the end of April. For consistency, I would be tempted to trim the data rows to a concise range from January 1 to March 31.

I'll sort it by date and Build the final product for now.

Build

The data is organized neatly, sorted correctly, and focused on what matters. The first two segments, Source and Focus, are linear. I returned to fix additional data, but reaching this stage was mostly straightforward. The final segment, Build, tends to be more iterative and experimental. It will take a few attempts to achieve exactly what I want.

For the first pass, I will collect monthly statistics. Hold on – the data is at a daily level, and I want to summarize it every month. I need to adjust the data once more. The date is in the correct format, but Excel interprets it as a string like "2023-04-24." I want Excel to recognize it as a date column.

CHAPTER 4 EXAMPLES – FILES

Data comes in three primary forms: numbers, text, and dates. While dates are essentially text, most platforms have methods for handling them that shouldn't be overlooked.

A quick fix using "DATEVALUE" looks like this in Figure 4-27.

	D	N
1	Date	Day Date
21	2023-04-24	=DATEVALUE(D21)
15	2023-04-23	

Figure 4-27. Carbon Dating

This results in a number. In Excel, a date represents the number of days since January 1, 1900. I'm almost there – format the cell as a date in Figure 4-28.

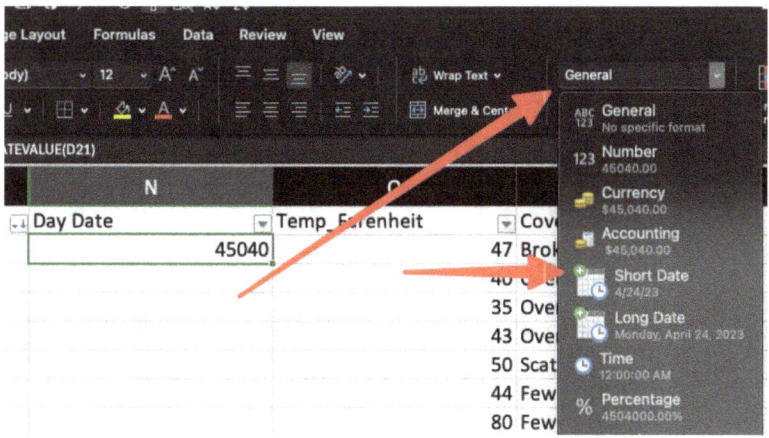

Figure 4-28. Formatting Fixes All

This results in a proper date (for my cultural locale). If I were in the United Kingdom or a part of Asia, these would match my expectations; Figure 4-29.

63

CHAPTER 4   EXAMPLES – FILES

D	N	
Date	Day Date	Te
2023-04-24	4/24/23	
2023-04-23		

Figure 4-29. *American-Style Dates*

Now that I've got the date sorted out, I'll hide the "D Date" column and make sure it's sorting the data by the date; Figure 4-30.

N	O		P
Day Date	Temp_Farenheit	Cover Name	
12/27/22	20	Broken	
12/28/22	35	Overcast	
12/29/22	35	Overcast	
12/30/22	21	Few Clouds	
12/31/22	30	Overcast	
1/5/23	26	Broken	
1/6/23	13	Few Clouds	
1/7/23	10	Clear	
1/8/23	18	Few Clouds	
1/9/23	24	Broken	
1/10/23	25	Overcast	

Figure 4-30. *More Usable*

So far, I've spent the first few minutes building, reviewing, and fixing the data. This is normal. Once the entire process is complete, I will include that date fix with the other corrections to ensure I'm ready when I reach the "Build" segment. This demonstrates the iterative nature of good analytics, which can get messy. Most importantly, since I'm following the Data Flow Map Framework, I have much more confidence in my data foundation and the overall process of managing the analysis.

CHAPTER 4   EXAMPLES – FILES

Before boxing up the data, I will copy only the useful columns from Figure 4-31 to a new worksheet. This represents a Ship, marking the end of the first phase of the analysis.

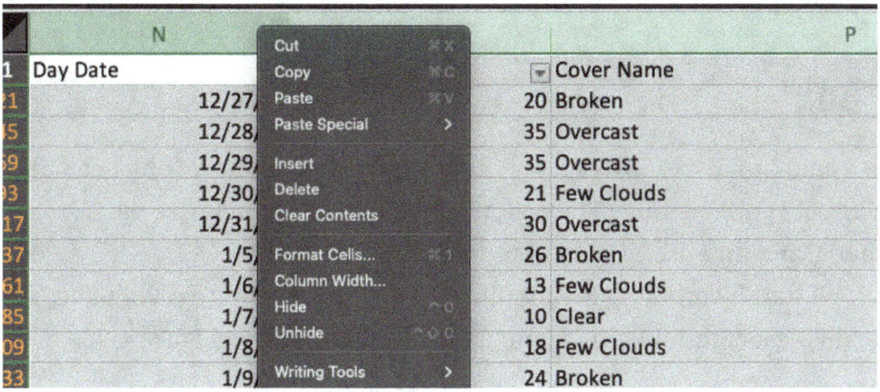

Figure 4-31. Copy for New Worksheet

By using the plus symbol on the right side of the existing worksheets, I will add a new Worksheet named "Date_Temp_Cover." Then I will paste the contents of the clipboard into that area. Refer to Figure 4-32.

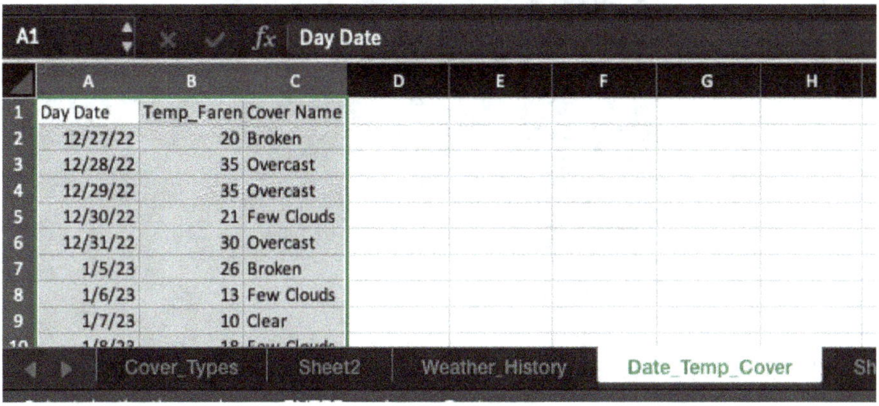

Figure 4-32. Paste into a New Worksheet

65

CHAPTER 4   EXAMPLES – FILES

Creating a new worksheet removes much of the unnecessary clutter from the source that I don't need. Now, I can start organizing the content. With Excel, building data occurs quickly and iteratively, yielding immediate results.

] Box

In traditional Excel, boxes refer to PivotTables, which have long been one of Excel's core features. They allow you to quickly summarize data, create charts, and find useful patterns with just a few clicks.

I'll start by profiling the monthly temperatures using a pivot table. To do this, select the "Insert" menu and "PivotTable," as in Figure 4-33.

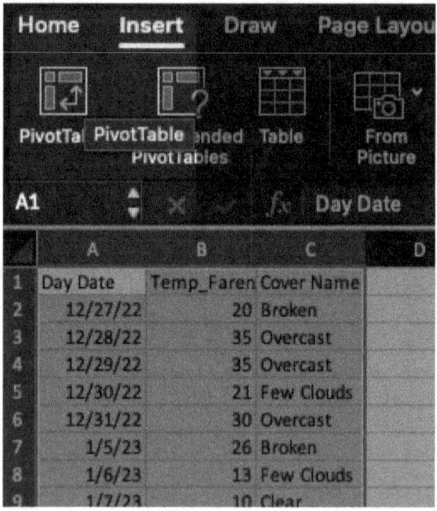

Figure 4-33. Start a New Pivot

I'll start with the basics and create a new worksheet; Figure 4-34.

CHAPTER 4   EXAMPLES – FILES

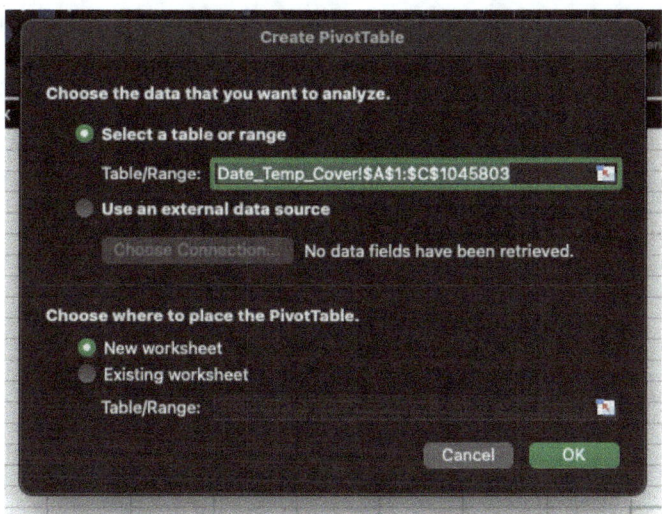

Figure 4-34. *Create PivotTable Dialog*

This results in a new worksheet with the blank Pivot canvas; Figure 4-35.

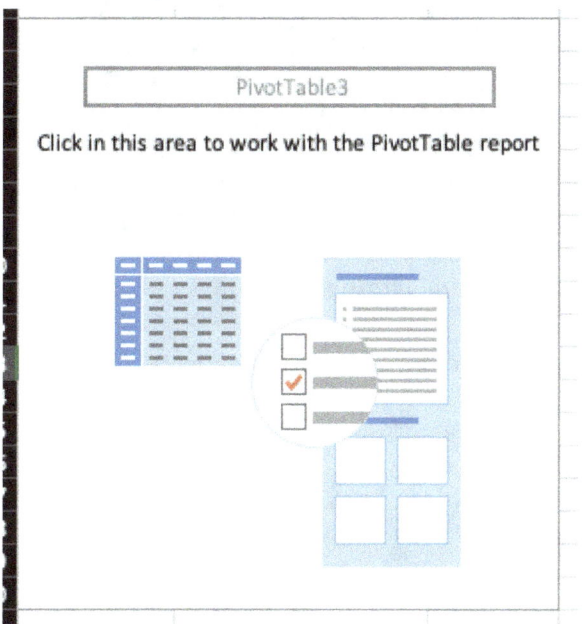

Figure 4-35. *Blank Pivot Canvas*

CHAPTER 4   EXAMPLES – FILES

The Pivot dialog on the right side of the screen makes it easy to start dragging and dropping content onto the PivotTable; Figure 4-36.

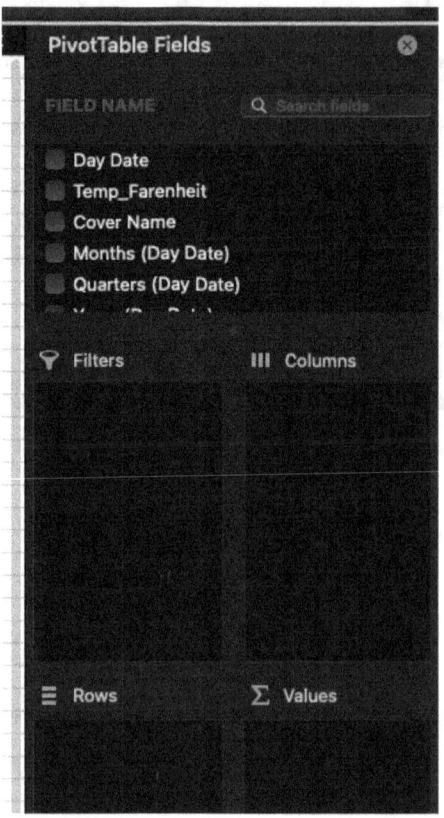

Figure 4-36. *PivotTable Dialog*

I'll drag the Day Date to "Rows" for the first Box; see Figure 4-37.

68

CHAPTER 4   EXAMPLES – FILES

3	Row Labels
4	⊟ 2022
5	⊟ Qtr4
6	⊟ Dec
7	12/27/22
8	12/28/22
9	12/29/22
10	12/30/22
11	12/31/22
12	⊟ 2023
13	⊟ Qtr1
14	⊟ Jan
15	1/5/23
16	1/6/23
17	1/7/23
18	1/8/23
19	1/9/23

Figure 4-37. First Pivot Table Values

I've expanded the dates to the day level. Since I want the Year and Month, I'll drag off the parts of the "Rows" field that I don't wish to keep; Figure 4-38.

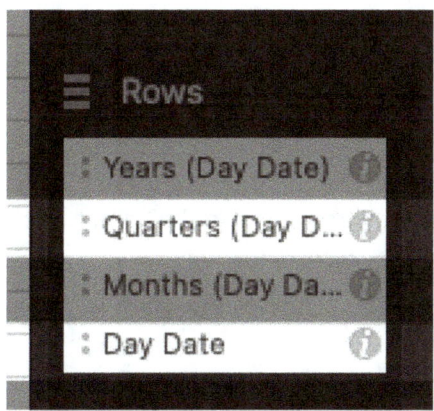

Figure 4-38. Things to Ignore

69

CHAPTER 4   EXAMPLES – FILES

As a result, I've boxed the data in a way that will be useful for the next steps; Figure 4-39.

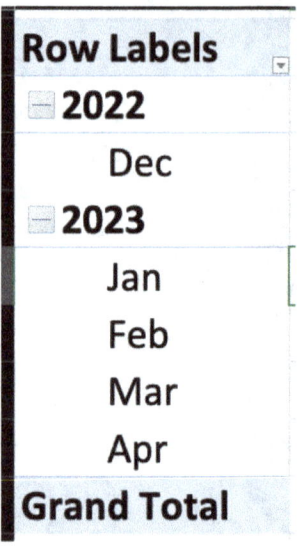

Figure 4-39. Just the Years and Months

Now that I've got boxes, the year, and the month, I can start digging into the data metrics in the following action: Size.

Size

Sizing is measuring impact. In this case, I will put a couple of metrics on this Pivot Table. Since Excel is speedy and visual with small datasets, I can iterate through this very quickly.

I'll drop the Temp_Fahrenheit into the Values box, and by default, it gives me the sum; Figure 4-40.

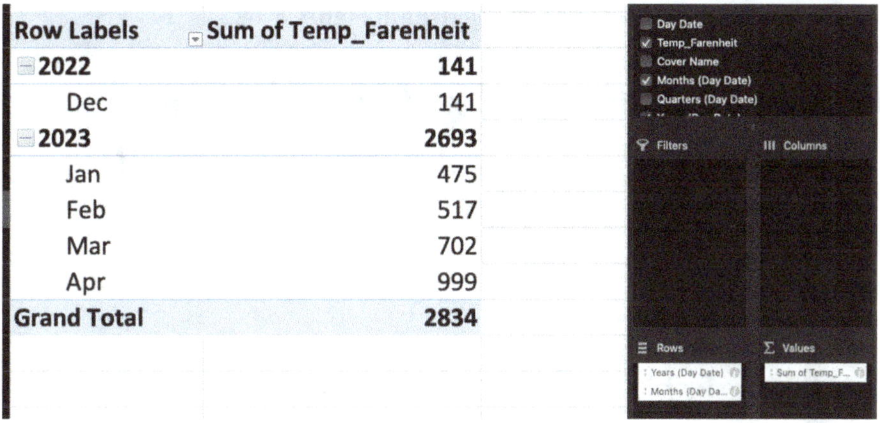

CHAPTER 4   EXAMPLES – FILES

Figure 4-40. *A Not Very Useful Metric*

Summing doesn't quite fit here, so I'm going to adjust the field by clicking the "I" button and changing the measure to "MAX," as shown in Figure 4-41.

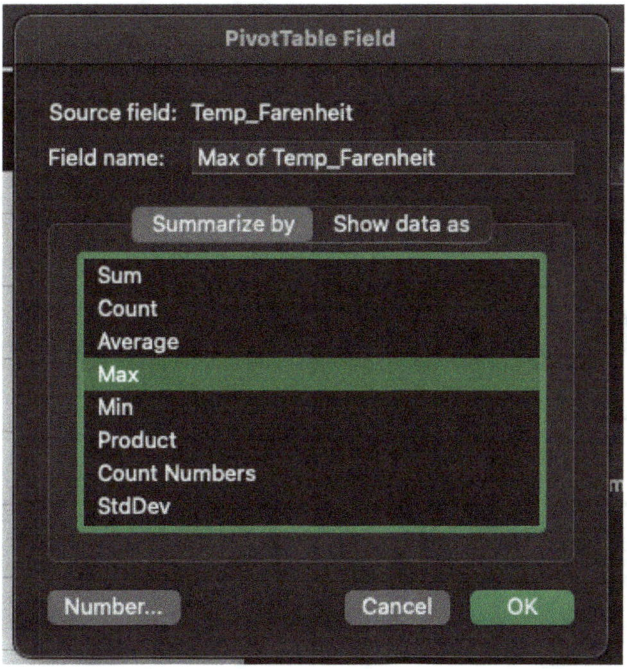

Figure 4-41. *Now We're Getting Somewhere*

71

CHAPTER 4   EXAMPLES – FILES

This looks more like something useful; Figure 4-42.

Row Labels	Max of Temp_Farenheit
⊟ 2022	35
Dec	35
⊟ 2023	80
Jan	35
Feb	39
Mar	43
Apr	80
Grand Total	80

Figure 4-42. *Better Metrics – Max*

Notice I'm getting "Totals" still. I'll use the PivotTable Design to turn Totals off, as seen in Figures 4-43, 4-44, 4-45 and 4-46.

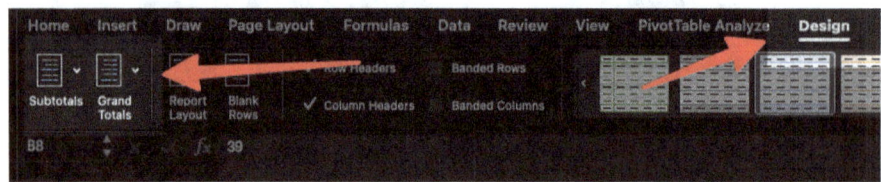

Figure 4-43. *Totals Sub and Grand*

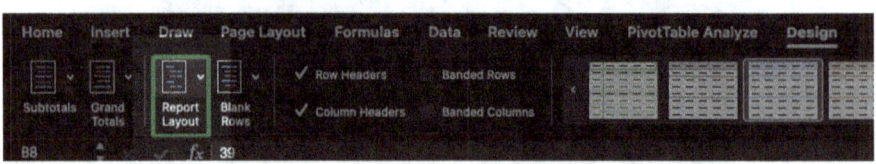

Figure 4-44. *Report Layout*

72

CHAPTER 4 EXAMPLES – FILES

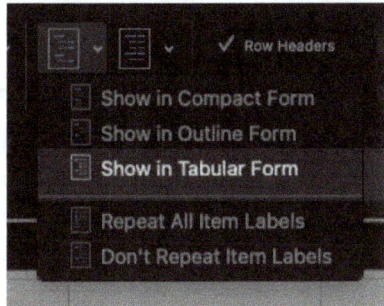

Figure 4-45. Report Layout Tabular

Years (Day D)	Months (Day Date)	Max of Temp_Farenheit
2022	Dec	35
2023	Jan	35
	Feb	39
	Mar	43
	Apr	80

Figure 4-46. Simple Layout

This is starting to come together. I can see the high temperature. Let's add the low and the average. I'll drag the temperature to the values box and edit the formula using the "I" button, see Figure 4-47.

Years (Day D)	Months (Day Date)	Min of Temp_Farenheit	Max of Temp_Farenheit	Average of Temp_Farenheit
2022	Dec	20	35	28.2
2023	Jan	2	35	19.79166667
	Feb	-5	39	22.47826087
	Mar	21	43	33.42857143
	Apr	33	80	52.57894737

Figure 4-47. Min, Max, and Average

This is almost looking ready to ship.

73

CHAPTER 4   EXAMPLES – FILES

= Ship

This is the last stage of the analysis, but it is almost certainly not the final iteration. After going through it, I can see a dozen ways to add more context, focus on sharper details, and build results that can make a difference.

My goal for this analysis is to create a monthly profile of the temperatures. I can easily generate bar, line, or scatterplot charts that effectively convey the story. However, for now, I will do some final cleanup before presenting the data as it is.

- Format the Average column as numbers and reduce the number of decimal points to zero.
- Change column headings that contain additional details from the analysis process.
- Repeat the year in the year column just to be more explicit.
- Update style of the table can be a little nicer. Excel has a lot of options.
- Cloud coverage average using the Pivot value options.
- Add a number count of the observations by dragging the date to the metric column.
- Relabel all of the columns to be more readable and center the text.

See Figure 4-48 for the reveal.

CHAPTER 4   EXAMPLES – FILES

Year	Month	Low	High	Average	Cloud Coverage	Observations
2022	Dec	20	35	28	83%	5
2023	Jan	2	35	20	75%	24
2023	Feb	-5	39	22	57%	23
2023	Mar	21	43	33	64%	21
2023	Apr	33	80	53	59%	19

Figure 4-48. *Final (For Now) Version*

Summary

Part of the analytics process is sharpening the perspective on the results. The data above isn't much different from the previous version; it's now much more consumable. Just glancing at this, I can tell:

- December is short on observations – is that a data quality problem? Should I ignore December altogether?

- January and February are very similar, with February slightly colder and warmer but generally heavy coat weather.

- March is progressively different, and April busts out of the cold weather pattern with a very warm 80-degree high.

- Skies get significantly bluer and less cloudy the nearer you get to spring.

75

CHAPTER 4   EXAMPLES – FILES

This analysis covered all the bases; see Table 4-1.

Table 4-1. *All of the modes and actions*

Mode	Symbol	Action	Goal
Source	O	Get	Gather raw Weather Observations
	+	Mix	Add in cloudiness index
	X	Fix	Fix the dates, convert temps from C to F
Focus	V	Cut	Just select date, temp and coverage
	>	Slice	Just select the noontime observation
	\	Sort	Order by date
Build	]	Box	Group by month
	#	Size	Temp - low, high and average, Coverage, Counts
	=	Ship	Clean up the table and simplify for publication

In the next chapter, I present database examples.

CHAPTER 5

Examples – Databases – SQL

Databases

Databases are the core of operations, analytics, and data science, among other areas in business and academia. While this chapter and book mainly focus on traditional SQL database systems, the Data Flow Map Framework will enhance creativity, communication, and clarity in nearly any organized data system.

Most databases are relational SQL-based systems. These include both open-source and proprietary options, such as PostgreSQL, Oracle, MySQL, DuckDB, Microsoft SQL Server, SQLite, and many other variants. Most databases support the ANSI Standard SQL, often with platform-specific variations that make them better suited for particular tasks. Database operations generally serve as a bridge between raw data and final destinations.

This book will not attempt to cover all features that each variation of a database offers, but instead show how basic database operations fit into the overall Data Flow Map Framework. For simplicity, I'll focus on DuckDB, a popular analytics database. At the time of writing this book,

CHAPTER 5   EXAMPLES – DATABASES – SQL

DuckDB offers a convenient web UI that facilitates easy execution of SQL queries against its database format.

My goal with this analysis is to examine personal coffee-purchasing data, identify patterns, and see if I prefer buying coffee from a local, single-owner shop over the usual Grande Mocha at a national chain. Let's see if the data aligns with my stated values.

The raw data is stored in CSV files that will be imported into the DuckDB database for the remainder of the analysis. Each action will result in a separate table, identified by its name. In a typical production database, you may not want to create multiple copies of tables, especially if the source data is extensive and complex. In this case, rely on views to maintain speed and clarity without sacrificing space.

Source

The initial datasets are in two CSV files:

- Sales data tracking a decade of purchases across diverse locations and amounts.

- Location data that indicates a company's size, ranging from a single local location to a global company present in nearly every airport.

The goal of this action is to load the data into the database, combine it, and clean it up.

o Get

DuckDB is highly flexible in terms of data sources. Opening a CSV file is almost as easy as finding it on disk, sketched out in Table 5-1. The code is simple – create a table (Listing 5-1), with the result shown in Figure 5-1. Each code block will include a comment with the action symbol and a quick tag.

CHAPTER 5   EXAMPLES – DATABASES – SQL

Table 5-1. *Sourcing Coffee Data*

@	**Get the sales**
o	Land raw sales data
+	Add in Location Category

Listing 5-1. Get Sales

```
--- o Get Sales
CREATE TABLE sales as
SELECT * FROM read_csv('./Coffee_Sales.csv');
```

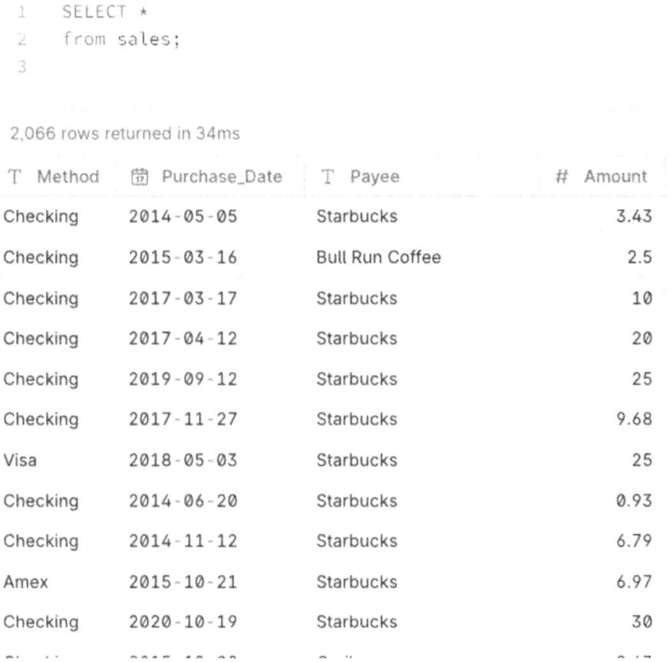

```
1  SELECT *
2  from sales;
3
```

2,066 rows returned in 34ms

T Method	📅 Purchase_Date	T Payee	# Amount
Checking	2014-05-05	Starbucks	3.43
Checking	2015-03-16	Bull Run Coffee	2.5
Checking	2017-03-17	Starbucks	10
Checking	2017-04-12	Starbucks	20
Checking	2019-09-12	Starbucks	25
Checking	2017-11-27	Starbucks	9.68
Visa	2018-05-03	Starbucks	25
Checking	2014-06-20	Starbucks	0.93
Checking	2014-11-12	Starbucks	6.79
Amex	2015-10-21	Starbucks	6.97
Checking	2020-10-19	Starbucks	30

Figure 5-1. *Get Sales Result*

CHAPTER 5 EXAMPLES – DATABASES – SQL

I will also want to get location data to mix in the next step, so I'll repeat this code (Listing 5-2), and then Mix them together.

Listing 5-2. Load Up the Locations

```
--- o Get Locations
CREATE TABLE locations
AS SELECT *
FROM read_csv('./Location_Category.csv');
```

This results in Figure 5-2.

```
1    SELECT * from locations;
```

49 rows returned in 16ms

T Store	123 Locations
Starbucks	40000
Caribou	750
Dunn Brothers Coffee	48
Urban Coffee	1
Sencha Tea Garden	3
Tea Garden	1

Figure 5-2. *Location Data Sampler*

+ Mix

Combining tables is a key strength of SQL, along with the ability to handle much higher row counts than Excel. I'll Mix the two tables with a join, as in Table 5-2.

CHAPTER 5   EXAMPLES – DATABASES – SQL

Table 5-2. Bringing It Together

@	Mix Together
0	Sales CSV File
x	Join Location on Payee

This is played out in the SQL, in Listing 5-3.

Listing 5-3. Join Tables

```
--- + Mix Sales and Location
CREATE TABLE mix as
SELECT *
FROM sales
join locations on sales.Payee = Locations.Store;
```

This gives the result in Figure 5-3.

```
1   SELECT * FROM mix
```

2,057 rows returned in 16ms

Method	Purchase_Date	Payee	Amount	Store	Locations
Checking	2014-01-06	Dunn Brothers Coffee	2.09	Dunn Brothers Coffee	48
Checking	2014-01-06	Dunn Brothers Coffee	2.09	Dunn Brothers Coffee	48
Checking	2014-01-06	Starbucks	10	Starbucks	40000
Checking	2014-01-06	Starbucks	10	Starbucks	40000
Checking	2014-01-06	Starbucks	3.43	Starbucks	40000
Checking	2014-01-06	Starbucks	1.84	Starbucks	40000
Checking	2014-01-07	Dunn Brothers Coffee	2.09	Dunn Brothers Coffee	48
Checking	2014-01-07	Starbucks	5	Starbucks	40000
Checking	2014-01-08	Starbucks	18.89	Starbucks	40000
Checking	2014-01-09	Dunn Brothers Coffee	4.64	Dunn Brothers Coffee	48
Checking	2014-01-09	Starbucks	3.43	Starbucks	40000
Checking	2014-01-10	Starbucks	5	Starbucks	40000
Checking	2014-01-10	Starbucks	5	Starbucks	40000
Checking	2014-01-10	Starbucks	0.33	Starbucks	40000

Figure 5-3. Joined at the hip Mix

CHAPTER 5 EXAMPLES – DATABASES – SQL

I want to be able to understand the data in different ways. I'll add a `Purchase_Year` and round the amount to the nearest dollar for later analysis; see Table 5-3.

Table 5-3. *Get Out the Toolbox*

@	Fixes
0	Mix : Sales and Size Joined
x	Pull year from purchase date as "Purchase_Year"
x	Round amount to the nearest decimal dropping pennies

Let's play this out in code with Listing 5-4.

Listing 5-4. Fixes in Place

```
--- x Fix Year and Amount
CREATE TABLE fix as
SELECT *,
  YEAR(Purchase_Date) as Purchase_Year,
  ROUND(Amount) as Amount_Dollar
FROM mix
```

Another look at the data shows the updates, in Figure 5-4.

CHAPTER 5   EXAMPLES – DATABASES – SQL

```
1   SELECT * FROM mix
```

2,057 rows returned in 45ms

T Method	🗓 Purchase_Date	T Payee	# Amount	T Store	T Category	123 Location Count
Checking	2014-05-05	Starbucks	3.43	Starbucks	National	40000
Checking	2015-03-16	Bull Run Coffee	2.5	Bull Run Coffee	Local	2
Checking	2017-03-17	Starbucks	10	Starbucks	National	40000
Checking	2017-04-12	Starbucks	20	Starbucks	National	40000
Checking	2019-09-12	Starbucks	25	Starbucks	National	40000
Checking	2017-11-27	Starbucks	9.68	Starbucks	National	40000
Visa	2018-05-03	Starbucks	25	Starbucks	National	40000
Checking	2014-06-20	Starbucks	0.93	Starbucks	National	40000
Checking	2014-11-12	Starbucks	6.79	Starbucks	National	40000
Amex	2015-10-21	Starbucks	6.97	Starbucks	National	40000

Figure 5-4. *Fixes in Place*

Focus

Now that I have a solid foundation for the data, I can begin focusing on the key parts with Focus. This is where SQL shines. It makes selecting columns, filtering rows, and sorting results much easier.

v Cut

After mixing in other tables, I began collecting columns as if they were small, rare collectibles on an auction site. I want to select, mapped out in Table 5-4.

> Payee
>
> Purchase Year
>
> Size Category

CHAPTER 5   EXAMPLES – DATABASES – SQL

Table 5-4. Just the Basics, Please

@	Focus
o	Mixed sales with fixes
v	Choose the Year, Category, and Amount
>	Only choose local stores
\	Sort by year

Listing 5-5. Clean Shot in SQL

```
--- v Cut down columns
CREATE TABLE cut as
SELECT Payee,
  Purchase_Date,
  Size_Category,
  Amount_Dollar
FROM fix
```

CHAPTER 5   EXAMPLES – DATABASES – SQL

This results in Figure 5-5.

```sql
1   SELECT * FROM cut;
```

2,057 rows returned in 29ms

T Payee	📅 Purchase_Date	T Size_Category	# Amount_Dollar
Dunn Brothers Coffee	2014-01-06	Area	2
Dunn Brothers Coffee	2014-01-06	Area	2
Starbucks	2014-01-06	National	10
Starbucks	2014-01-06	National	10
Starbucks	2014-01-06	National	3
Starbucks	2014-01-06	National	2
Dunn Brothers Coffee	2014-01-07	Area	2
Starbucks	2014-01-07	National	5
Starbucks	2014-01-08	National	19

Figure 5-5. *Narrower Focus*

> Slice

While I love my Buck's, let's focus on the local shops; Table 5-5.

Table 5-5. *Focus on what matters most*

@	Focus
o	Mixed Sales Data
v	Choose the Year, Category, and Amount
>	Only choose local stores

85

CHAPTER 5   EXAMPLES – DATABASES – SQL

Using the code in Listing 5-6, reduce the rows.

Listing 5-6. Shorter and Sweeter

```
--- > Slice to Local
CREATE or REPLACE  TABLE slice as
SELECT *
FROM cut
WHERE Category = 'Local'
```

This results in Figure 5-6.

```
1    Select * FROM Slice;
```

113 rows returned in 15ms

Payee	Purchase_Year	Category	Amount_Dollar
Bull Run Coffee	2015	Local	3
Urban Coffee	2022	Local	5
Urban Coffee	2022	Local	10
Urban Coffee	2022	Local	29
Tea Garden	2016	Local	16
Urban Coffee	2022	Local	10
Urban Coffee	2022	Local	13
True Stone	2020	Local	5
Urban Coffee	2022	Local	13

Figure 5-6. *Fewer Rows*

= Sort

These are randomly ordered; let's sort according to the Purchase_Year (Table 5-6).

Table 5-6. *Rank order by year*

@	Focus
o	Mixed Sales Data
v	Choose the Year, Category, and Amount
>	Only choose local stores
\	Sort by Year

The data is sorted by Purchase_Year, as in Listing 5-7.

Listing 5-7. Sort by Year

```
--- \ Sort by Year
CREATE TABLE sort as
SELECT *
FROM slice
ORDER BY Purchase_Year
```

CHAPTER 5   EXAMPLES – DATABASES – SQL

This results in Figure 5-7.

```
1   SELECT * FROM sort
```

113 rows returned in 25ms

T Payee	123 Purchase_Year	T Category	# Amount_Dollar
Black Dog Coffee	2014	Local	6
Tea Garden	2014	Local	10
Bull Run Coffee	2014	Local	4
Nokomis Beach Coffee	2014	Local	2
Tea Garden	2014	Local	6
Tea Garden	2014	Local	4
Tea Garden	2014	Local	5
I am Coffee	2014	Local	4
I am Coffee	2014	Local	4
Tea Garden	2014	Local	4
Overlook Coffee	2014	Local	2

Figure 5-7. *Ordered by Year*

Build

I've got my foundational data lined up and focused on just the key pieces of information. Now it's time to roll it up to make a point about how often I actually shop local. By doing this, I can create something new. I'll start with summarizing with Box, calculating impact with Size, and finally delivering a result using Ship.

] Box

I'll group by Payee and Year; Table 5-7.

Table 5-7. *Start bucketing*

@	Build
]	Group by Payee and Year

Group up the data as in Listing 5-8.

Listing 5-8. Boxing It Up to Go

```
--- ] Box By Payee and Year
CREATE TABLE Box AS
SELECT Payee,
  Purchase_Year
from sort
GROUP BY
Payee,
Purchase_Year
```

CHAPTER 5   EXAMPLES – DATABASES – SQL

This results in Figure 5-8.

```
1    SELECT * from box;
```

42 rows returned in 27ms

Payee	Purchase_Year
Zoe Coffee	2019
Java Moose	2018
COFFEE PAW CAFE	2021
Tea Garden	2015
Corner Coffee	2020
Moose And Sadies	2016

Figure 5-8. Simpler, Clearer

Size

This has been reduced to 42 rows, which is much better, but I want to know the total amount and how many times per year I visited the local shops. In the process, I plan to combine a Box and Size into one step and add a row count for context, a common practice in SQL queries. This looks like Table 5-8.

Table 5-8. *How big is the effect?*

@	Build
]	Group by Payee and Year
#	Size by amount and row count

Calculate the effect, as in Listing 5-9.

Listing 5-9. Measure Impact

```
--- # Size by Year
CREATE TABLE size AS
SELECT Purchase_Year,
  SUM(Amount_Dollar) as Annual_Amount,
  COUNT(*) as Row_Count
FROM sort
GROUP BY Purchase_Year
```

CHAPTER 5   EXAMPLES – DATABASES – SQL

This results in Figure 5-9.

```
1    SELECT * FROM size
```

10 rows returned in 28ms

123 Purchase_Year	# Annual_Amount	123 Row_Count
2014	64	14
2015	73	16
2016	141	12
2017	49	5
2018	105	4
2019	89	6
2020	33	3
2021	73	5
2022	488	44
2023	48	4

Figure 5-9. *Box and Size in One Shot*

= Ship

Finally, I want to wrap this up into a CSV file to use a proper charting tool, for example, Table 5-9.

Table 5-9. *Ship it out the door*

@	Build
]	Group by Payee and Year
#	Size by amount and row count
=	Deliver final CSV file

Let's move it out in Listing 5-10.

Listing 5-10. Ship It Out the Door

```
--- = Ship to back to CSV
COPY size to yearly_local_volume.csv (HEADER, DELIMITER ',');
```

The file opened in a text editor outside of the database, looks like Figure 5-10.

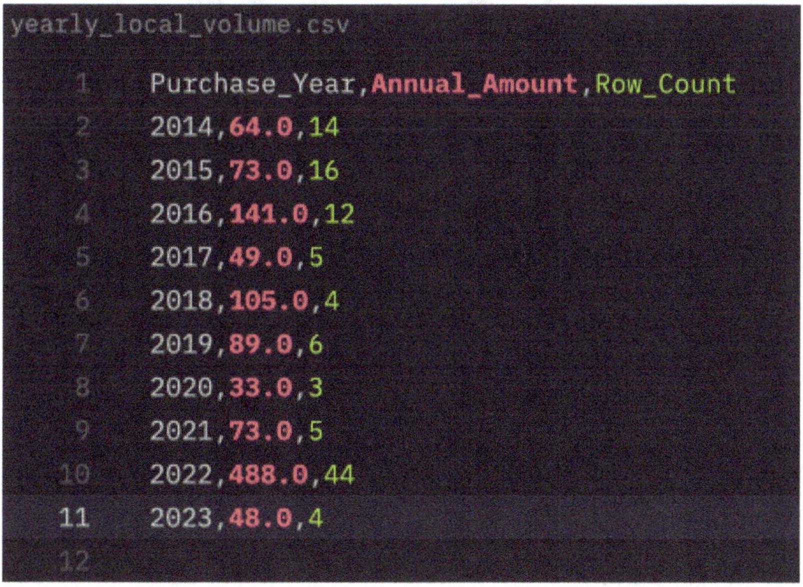

```
yearly_local_volume.csv
 1    Purchase_Year,Annual_Amount,Row_Count
 2    2014,64.0,14
 3    2015,73.0,16
 4    2016,141.0,12
 5    2017,49.0,5
 6    2018,105.0,4
 7    2019,89.0,6
 8    2020,33.0,3
 9    2021,73.0,5
10    2022,488.0,44
11    2023,48.0,4
12
```

Figure 5-10. *Final Results*

Summary

In this example, I transformed raw data into something that makes a good conversation starter. It's not finished, and it might not even be in the final format, but it's a solid beginning. I can easily update the code, knowing I've put these pieces together, as shown in Table 5-10.

CHAPTER 5   EXAMPLES – DATABASES – SQL

Table 5-10. *Overview*

Motion	Symbol	Action	Goal
Source	O	Get	Gathered coffee purchase data
	+	Mix	Add in location category
	X	Fix	Converted
Focus	V	Cut	Just Useful Columns
	>	Slice	Just Useful Data
	\	Sort	Ordered by Year
Build	]	Box	Group by Payee Year
	#	Size	Counted Events
	=	Ship	Clean up to CSV

When I annotate SQL code with the Data Flow Map Framework, I add comments at the beginning of each section. If I remove the actual code from all previous sections, I am left with a descriptive series of DFM actions, as seen in Listing 5-11.

Listing 5-11. Comments Only Please

```
--- o Get Locations
--- + Mix Sales and Location
--- x Fix Year and Amount
--- v Cut down columns
--- > Slice to Local
--- \ Sort by Year
--- ] Box By Payee and Year
--- # Size by Year
--- = Ship to back to CSV
```

CHAPTER 5   EXAMPLES – DATABASES – SQL

The Data Flow Map Framework is flexible enough to map out text-based systems, such as SQL code, and break them into manageable parts, making it easier to understand than hundreds of lines of confusing code.

SQL code and databases can handle and store a vast amount of data. However, if I don't understand the process clearly, it becomes difficult to explain it to others and, more importantly, to be creative without risking critical analytics being messed up. The Data Flow Map can help with all of those challenges.

In the next chapter, I present Python examples.

CHAPTER 6

Examples – Python

Python Code

This chapter will use Python to process flight data through the Data Flow Map. It will focus on a specific type of flight and build a result that tells us how often this type of flight passes over our location and in what directions.

For my purposes, I will focus on the types of flights that produce contrails in the atmosphere. Airplanes flying at altitudes between 30,000 and 40,000 feet create contrails. These contrails form when the hot exhaust from the aircraft engines mixes with the cold air at these high altitudes, forming ice crystals. Consequently, these high-altitude flights often leave visible trails in the sky, which can be tracked and analyzed, such as utilizing a Raspberry Pi, an antenna, and software to collect and interpret flight data. The goal of this analysis will be to compare the number of flights by direction to the airline. To get there, I'll use the following big motions:

- **Source**: Pull in-flight tracking data, mix in airframe details, create a conversion for track to direction.

- **Focus**: Select only flights above 30,000 feet, incorporating direction (track).

- **Build**: Relate the direction of the aircraft to the airline.

CHAPTER 6 EXAMPLES – PYTHON

Python has several useful utilities for sourcing and managing data; I will use Pandas, the most popular option. As the data is in Parquet format, I will use the PyArrow library. Throughout this example, I will list the code along with screenshots of the results.

At the head of the code are the following imports; Listing 6-1.

Listing 6-1. Python Library Imports

```
import pandas as pd
import pyarrow
```

Source

This example illustrates how to manipulate data using Python. The data will be provided in Parquet, a highly portable file format. Most of the effort will involve joining and fixing the data to make it useful.

o Get

The Flights data has been gathered from the Raspberry Pi and collected in a file called called Flights.parquet. For this exercise, I'll be using Parquet rather than CSV - it stores the metadata and data in one spot, making it easier to consume in a process like this one in Python, see Listing 6-2.

Listing 6-2. Loading in the Flight Data into a Data Frame

```
df_flights = pd.read_parquet("./Flights.parquet")
df_flights.head()
```

The head() function will show the first few rows and columns in a readable format; Figure 6-1.

	icao24	flight	ts_date	ts_time	alt	track	groundspeed	af_category	registration
0	06a073	QTR99V	2023-10-11	08:05:33	38000	192	488	A5	A7-BBC
1	06a073	QTR99V	2023-10-28	08:14:04	36000	194	393	A5	A7-BBC
2	06a073	None	2023-10-30	08:00:34	36000	190	454	A5	A7-BBC
3	06a07d	QTR99V	2023-10-19	07:47:31	38000	190	539	A5	A7-BBG
4	06a07d	None	2023-11-17	07:38:56	38000	189	504	A5	A7-BBG

Figure 6-1. *First Glance at the Data*

For clarity, I'm going to show one record vertically, using Listing 6-3.

Listing 6-3. Show Me One Record

```
one_record = df_flights.iloc[0]
print(one_record)
```

This gives me a complete listing of all of the fields; Figure 6-2.

CHAPTER 6 EXAMPLES – PYTHON

```
>>> one_record = df_flights.iloc[0]
>>> print(one_record)
icao24                               06a073
flight                               QTR99V
ts_date                 2023-10-11 00:00:00
ts_time                            08:05:33
alt                                   38000
track                                   192
groundspeed                             488
af_category                              A5
registration                         A7-BBC
manufacturername                     Boeing
model                             777 2DZLR
typecode                               B77L
operator                       Qatar Airways
operatorcallsign                    QATARI
operatoricao                            QTR
owner_name                     Qatar Airways
built                   2009-01-01 00:00:00
Name: 0, dtype: object
```

Figure 6-2. *All the Fields at Once*

+ Mix

The data looks good initially, but I want to incorporate more information about the airplane's size. I have a theory that this might be interesting in the final analysis. The af_category field provides a code; I want to use a lookup table called Aircraft_Category; Listing 6-4.

Listing 6-4. Load Aircraft Category

```
# Aircraft Category
df_aircraft_category = pd.read_parquet("./Aircraft_Category.parquet")
df_aircraft_category.head()
```

This gives Figure 6-3.

	Category	Description
0	A1	Light (< 15 500 lbs.)
1	A2	Small (15 500 to 75 000 lbs.)
2	A3	Large (75 000 to 300 000 lbs.)
3	A4	High Vortex Large(aircraft such as B-757)
4	A5	Heavy (> 300 000 lbs.)

Figure 6-3. Aircraft Category

That looks great, but I want to incorporate the "Description" field in the primary source Flights table. This can be done with a merge; Listing 6-5.

Listing 6-5. Merge Dataframes

```
merged_df = pd.merge(df_flights, df_aircraft_category, left_on='af_category', right_on='Category')
merged_df.head()
```

For the sake of illustration, I'm going to briefly focus on just the index, the af_category, and description in the results; Figure 6-4.

	icao24	af_category	Category	Description
0	06a073	A5	A5	Heavy (> 300 000 lbs.)
1	06a073	A5	A5	Heavy (> 300 000 lbs.)
2	06a073	A5	A5	Heavy (> 300 000 lbs.)
3	06a07d	A5	A5	Heavy (> 300 000 lbs.)
4	06a07d	A5	A5	Heavy (> 300 000 lbs.)

Figure 6-4. It's a Match

CHAPTER 6   EXAMPLES – PYTHON

x Fix

Now that I've joined the Description information, I realize that perhaps I've got too much information – maybe I reduce that to just the first word? Also, when I look at the "track" column indicating direction the flight is going, it's expressed as degrees on the compass. For simplicity's sake, I'd like to convert direction to the easier-to-process N, W, S, E, NW, NE, SE, and SW.

I'll convert the Description to a Short_Description using a lambda function in Listing 6-6.

Listing 6-6. Split Out the Short_Description

```
merged_df['Short_Description'] = merged_df['Description'].
apply(lambda x: x.split()[0])
merged_df[['Description', 'Short_Description']].head()
```

This results in Figure 6-5.

	Description	Short_Description
0	Heavy (> 300 000 lbs.)	Heavy
1	Heavy (> 300 000 lbs.)	Heavy
2	Heavy (> 300 000 lbs.)	Heavy
3	Heavy (> 300 000 lbs.)	Heavy
4	Heavy (> 300 000 lbs.)	Heavy

Figure 6-5. *Shorter Description*

Now I want to categorize the degrees of direction to compass directions, in other words, from 270 degrees to "W", like in a compass rose; Figure 6-6.

CHAPTER 6 EXAMPLES – PYTHON

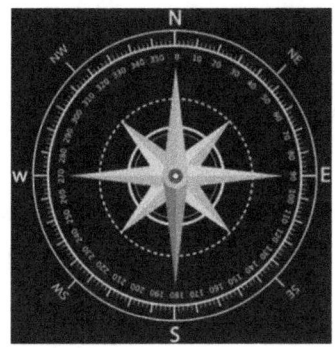

Figure 6-6. *Mapping Degrees to Directions*

Listing 6-7 will create a function for conversion.

Listing 6-7. Convert Degrees to Directions

```
# Convert compass degrees to cardinal direction
def compass_to_cardinal(compass):
    if compass > 337.5 or compass <= 22.5:
        return "N"
    elif compass > 22.5 and compass <= 67.5:
        return "NE"
    elif compass > 67.5 and compass <= 112.5:
        return "E"
    elif compass > 112.5 and compass <= 157.5:
        return "SE"
    elif compass > 157.5 and compass <= 202.5:
        return "S"
    elif compass > 202.5 and compass <= 247.5:
        return "SW"
    elif compass > 247.5 and compass <= 292.5:
        return "W"
```

103

```
    elif compass > 292.5 and compass <= 337.5:
        return "NW"
    else:
        return "Unknown"
```

I can now apply this to the data frame as a new column based on the track number, as shown in Listing 6-8.

Listing 6-8. Directionality

```
merged_df['Direction'] = merged_df['track'].apply(compass_to_
cardinal)
merged_df[['track', 'Direction']].head()
```

This gives this nicely formatted direction; Figure 6-7.

	track	Direction
0	192	S
1	194	S
2	190	S
3	190	S
4	189	S

Figure 6-7. *Numbers to Letters*

Finally, I want to calculate the airplane's age. The "built" date includes the year I need. With a little date manipulation, I can extract that information in Listing 6-9.

Listing 6-9. Extract Year

```
merged_df['year_built'] = merged_df['built'].dt.year
merged_df[['built', 'year_built']].head()
```

This gives just the year as expected in Figure 6-8.

	built	year_built
0	2009-01-01	2009
1	2009-01-01	2009
2	2009-01-01	2009
3	2010-01-01	2010
4	2010-01-01	2010

Figure 6-8. *Year to Set Up Age*

With the year in hand, I can add the age, using Listing 6-10. Note that as of the writing of this book, it's 2025. Your math will probably look slightly different in the future:

Listing 6-10. Calculate Age from Today

```
# Calculate the current year
current_year  = datetime.datetime.now().year
merged_df['plane_age'] = current_year - merged_df['year_built']
merged_df[['icao24', 'year_built', 'plane_age']].head()
```

And that's how old the plane is; Figure 6-9.

CHAPTER 6   EXAMPLES – PYTHON

	year_built	plane_age
0	2009	16
1	2009	16
2	2009	16
3	2010	15
4	2010	15

Figure 6-9. Simple Age

After tweaking the data, I'm ready to focus on what's truly important.

Focus

Source section, I've gathered flight data, incorporated descriptions, and added new columns for shorter descriptions and directions. Now, it's time to concentrate on the valuable details. There are a lot of extra columns in that data frame; see Figure 6-10.

```
icao24: 06a073
flight: QTR99V
ts_date: 2023-10-11 00:00:00
ts_time: 08:05:33
alt: 38000
track: 192
groundspeed: 488
af_category: A5
registration: A7-BBC
manufacturername: Boeing
model: 777 2DZLR
typecode: B77L
operator: Qatar Airways
operatorcallsign: QATARI
operatoricao: QTR
owner_name: Qatar Airways
built: 2009-01-01 00:00:00
Direction: S
Category: A5
Description:  Heavy (> 300 000 lbs.)
Short_Type_Code:  Heavy
Short_Description: Heavy
year_built: 2009
plane_age: 16
```

Figure 6-10. Too Much Magic

v Cut

First things first, let's collect the data columns that matter:

- **Alt**: Altitude so I can pick the high planes
- **Short_Description**: An easy handle for how big the plane is
- **Direction**: Where is it headed
- **Plane Age**: Are older planes more common than new planes?

This is easy to do in Pandas; Listing 6-11, which results in Figure 6-11.

Listing 6-11. Hone in on the Essential Columns

```
simple_flights =
    merged_df[['alt',
        'Short_Description',
        'Direction',
        'plane_age']]
simple_flights.head()
```

	alt	Short_Description	Direction	plane_age
0	38000	Heavy	S	16
1	36000	Heavy	S	16
2	36000	Heavy	S	16
3	38000	Heavy	S	15
4	38000	Heavy	S	15

Figure 6-11. *Just the Facts*

CHAPTER 6   EXAMPLES – PYTHON

> Slice

That's great – much less horizontal detail. But when I count the rows, I get over 1,800. Now, let's focus on flights above 35,000 feet, using Listing 6-12, resulting in Figure 6-12.

Listing 6-12. Filter Above 35,000 Feet

```
# Filter flights above 35,000 feet
filtered_flights = simple_flights[simple_flights['alt']
> 35000]
filtered_flights.head()
```

	icao24	flight	ts_date	ts_time	alt
0	06a073	QTR99V	2023-10-11	08:05:33	38000
1	06a073	QTR99V	2023-10-28	08:14:04	36000
2	06a073	None	2023-10-30	08:00:34	36000
3	06a07d	QTR99V	2023-10-19	07:47:31	38000
4	06a07d	None	2023-11-17	07:38:56	38000

Figure 6-12. *Flights Above 35,000 Feet, Only 670 Rows*

Another handy way to slice the data, especially in discovery mode, is to take a random sample. Most analytic platforms and tools provide a way of taking a random sample that will show odd patterns and, most importantly, avoid some of your own bias, which typically pops up in the next section, "Sort."

Sampling in Pandas is very straightforward using the `sample(n=y)` function; Listing 6-13.

CHAPTER 6　EXAMPLES – PYTHON

Listing 6-13. Short, Sweet Sample

```
# Get a random sample of 10 records from simple_flights
random_sample = simple_flights.sample(n=10)
random_sample.head(10)
```

Figure 6-13 shows some interesting new data that wouldn't have surfaced otherwise. For example, it appears there are a lot of "heavy" flights at this altitude – is there a correlation with direction? I'm going to apply sampling to all the work ahead.

	alt	Short_Description	Direction	plane_age
114	31975	Heavy	NW	25
1225	33000	Heavy	NE	12
270	34000	Large	SW	4
1397	33000	Heavy	SE	13
735	35000	Heavy	SE	25
1017	37000	Heavy	SE	33
885	30025	Large	NW	3
1229	36000	Heavy	N	33
1015	36000	Heavy	NW	32
471	36975	Heavy	E	5

Figure 6-13. *Almost Entirely Random*

\ Sort

I know that the bottom altitude is now 35,000 feet. I'm curious – what are the upper values? I'll sort by the altitude descending; Listing 6-14.

109

CHAPTER 6　EXAMPLES – PYTHON

Listing 6-14. Sort by Altitude, Top to Bottom

```
sorted_flights =
     filtered_flights.sort_values('alt', ascending=False)
sorted_flights[['flight', 'alt', 'Direction']].head()
```

That's interesting – I didn't expect to be capturing flights above 100,000 feet. See Figure 6-14.

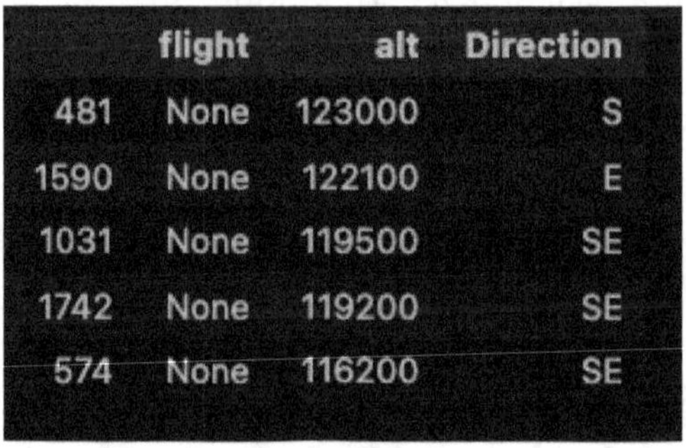

	flight	alt	Direction
481	None	123000	S
1590	None	122100	E
1031	None	119500	SE
1742	None	119200	SE
574	None	116200	SE

Figure 6-14. *Very High Aircraft Indeed*

A little background research on the aircraft model at the top showed that it can reach a maximum altitude of 41,000 feet. I have some incorrect data in the mix, so I will adjust the slice to include only flights between 35,000 and 41,000 feet. I will modify the code in Listing 6-15.

Listing 6-15. Range Altitude Between 35,000 and 41,000 Feet

```
# Slice the data more tightly
filtered_flights =
     simple_flights[(simple_flights['alt'] >= 35000) &
     (simple_flights['alt'] <= 41000)]
# Resort the flights
sorted_flights = filtered_flights.sort_values('alt',
ascending=False)
sorted_flights.head()
```

Let's take a look at what's left, in Figure 6-15.

	alt	Short_Description	Direction	plane_age
741	41000	Heavy	NW	23
857	41000	Heavy	E	7
667	41000	Light	SE	46
898	41000	Heavy	E	6
727	41000	Heavy	NW	21

Figure 6-15. Looking More Likely

A couple of thoughts on this exercise. First, why are there several records with altitude above 100,000 feet? Aside from military jets, there isn't much that typically flies at that altitude. Since this occurs in more than one record, I'm curious if there's a pattern with other parts of the data – the airline, the type of plane, and the preceding and following records for that flight. What happened? There's a lot of opportunity to dig deeper, and while this data may not matter, perhaps your data for work or home shows odd patterns.

While building an analysis, you will find yourself in dead ends, strange circles, and cul-de-sacs that don't help. It's okay to go back and rebuild the sourcing and focusing components and shift fixes back to the front of the bus. Have some data that isn't helpful? Slice it out. Do you have other pieces that aren't aligned correctly, or do you need one part of a field? Fix it. As you build new structures, you may find that the foundations don't work. Be ready to correct the flow – you have the confidence now that it won't completely derail your train of thought, and it will be better.

The following section will start to dig into those patterns.

CHAPTER 6   EXAMPLES – PYTHON

Build

I've got a solid source and am focused on the important. I will create something amazing and new with these raw materials. I'll start by boxing up categories, then move on to sizing the impact, and finally ship a completed project.

] Box

A box in Pandas is typically a "groupby", either with one column or several. You can also:

- Create pivot tables using the pivot_table() function.
- Slice values by quantiles using the qcut() function
- Build frequency analysis using crosstab().

I'll focus on just a simple "groupby" for now, on the direction. If I apply it, as in Listing 6-16, I'll get a simple result.

Listing 6-16. Basic Result

```
# Categorize by Direction
categorized_flights = filtered_flights.groupby('Direction')
```

This results in just Figure 6-16.

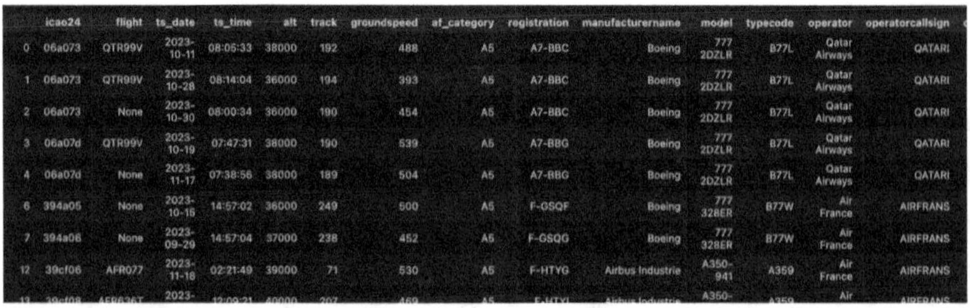

Figure 6-16. *Still a Lot More Than I Hoped For*

CHAPTER 6   EXAMPLES – PYTHON

If I want to narrow it down to just the key group by field, I need to consider metrics – size, which is next.

Size

Pandas and its associated NumPy library provide you with the entire universe of measurement tools. Once I've blocked something using a box, it's easy to create simple metrics. The simplest approach at first is often the best – the count of rows. I would like to see if one direction stands out more than another, as seen in Listing 6-17.

Listing 6-17. Count Rows by Direction

```
# Count the number of flights in each direction
flight_counts = filtered_flights.groupby('Direction').size()
flight_counts.head()
```

As Figure 6-17 shows, most flights go east and northwest.

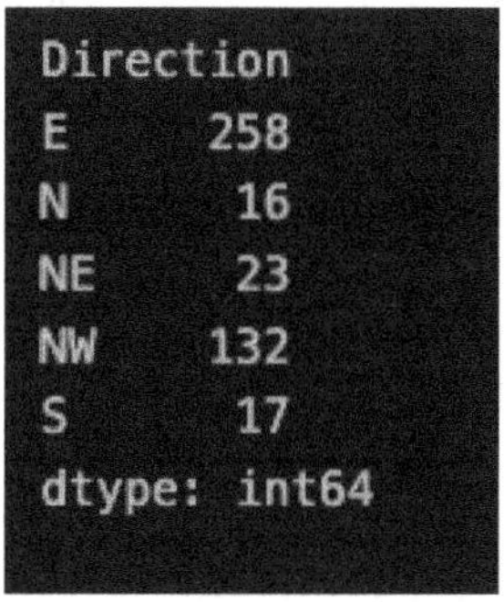

```
Direction
E      258
N       16
NE      23
NW     132
S       17
dtype: int64
```

Figure 6-17. Count by Direction

The original goal of this analysis is to compare the number of flights by direction to the airline. Now that I know the data is flowing correctly, I can create a pivot table and compare the two. This will be a step back into a Box, but that's okay. You iterate because you can do it safely and quickly.

CHAPTER 6   EXAMPLES – PYTHON

The code for this will be much more complex compared to the previous iterations; Listing 6-18, result in Figure 6-18.

Listing 6-18. Complete PivotTable

```
# Create pivot table
direction_operator_pivot = pd.pivot_table(
    filtered_flights,
    values='flight',
    index='operator',
    columns='Direction',
    aggfunc='count',
    fill_value=0
)

# Sort by total number of flights
direction_operator_pivot['Total'] = direction_operator_pivot.sum(axis=1)
direction_operator_pivot = direction_operator_pivot.sort_values('Total', ascending=False)

# Display top airlines
direction_operator_pivot.head(10)
```

This gives the results in a tidy table.

CHAPTER 6   EXAMPLES – PYTHON

Direction	E	N	NE	NW	S	SE	SW	W	Total
operator									
Delta Air Lines	42	0	5	5	0	20	2	14	88
Alaska Airlines	15	0	0	1	0	17	1	4	38
Federal Express	0	7	0	9	0	18	0	1	35
United Airlines	9	0	1	1	0	7	2	9	29
United Parcel Service	0	0	0	9	0	17	0	0	26
Kalitta Air	0	0	0	3	0	10	0	1	14
Atlas Air	1	0	0	2	1	8	0	0	12
American Airlines	2	0	0	1	0	2	2	4	11
British Airways	5	0	3	0	1	0	1	0	10
Polar Air Cargo	0	0	0	3	0	6	0	0	9

Figure 6-18. *Totals by Operator and Then Direction*

= Ship

I'm almost there, but I think there are simple things I could do to make the data pop more effectively. The final stage is the tipping point, and I often go back and forth several times before finalizing a result. The result in Figure 6-18 above is acceptable, but I have a bunch of zeros that make it a bit harder to read.

I'll replace them with a dot in Listing 6-19.

Listing 6-19. Simple Formatting

```
# Replace 0 values with '.' in the pivot table
direction_operator_pivot = direction_operator_pivot.replace(0, '.')
direction_operator_pivot.head(10)
```

CHAPTER 6   EXAMPLES – PYTHON

This results in Figure 6-19.

Direction operator	E	N	NE	NW	S	SE	SW	W	Total
Delta Air Lines	42	.	5	5	.	20	2	14	88
Alaska Airlines	15	.	.	1	.	17	1	4	38
Federal Express	.	7	.	9	.	18	.	1	35
United Airlines	9	.	1	1	.	7	2	9	29
United Parcel Service	.	.	.	9	.	17	.	.	26
Kalitta Air	.	.	.	3	.	10	.	1	14
Atlas Air	1	.	.	2	1	8	.	.	12
American Airlines	2	.	.	1	.	2	2	4	11
British Airways	5	.	3	.	1	.	1	.	10
Polar Air Cargo	.	.	.	3	.	6	.	.	9

Figure 6-19. *Cleaner Output*

This is a lot better.

Summary

Right off the bat, I can see a big difference between the local airline (Delta) and the others. It also seems like NW and SE are the primary directions. At this point, I'd like to revisit the data and determine if there's a way to capture origin and destination – what directions justify this traffic. Finally, notice that the passenger airlines (Delta, United, Alaska) have significantly more E and W flights than the cargo planes (Kalitta, Atlas, Polar), and the delivery airlines (FedEx, UPS) follow a specific pattern due to their spoke-and-hub distribution approach.

In this example, I sourced size data (which never actually surfaced), fixed the data direction, ultimately focused only on the airline operator and direction, and shipped a straightforward result that can serve as a conversation starter for more engaging analytics, some of which may be dull and uninteresting. In contrast, others could genuinely change your world.

CHAPTER 6   EXAMPLES – PYTHON

Table 6-1 lays out the big motions and actions.

Table 6-1. Overview

Motion	Symbol	Action	Goal
Source	O	Get	Gathered the flight tracking data
	+	Mix	Add in size data
	X	Fix	Converted compass direction
Focus	V	Cut	Just airline, direction, age and description
	>	Slice	Just flights in a narrow band of altitudes, sampled
	\	Sort	Ordered by altitude for testing
Build	]	Box	Group by airline and direction
	#	Size	Counted total number of records
	=	Ship	Clean up the table and simplify for publication

In my case, I now realize that there are three distinct types of airlines creating contrails above me:

- Passengers can be in almost any direction but have a significant east-west component.

- Cargo jets typically arc from Alaska down to Chicago in an NW/SE track.

- Delivery jets all go into one location – their company's hub.

That's already three separate analyses and plenty more where that came from. I could take the age data by airline and see who flies the oldest and newest planes or has the broadest range of ages.

117

CHAPTER 6   EXAMPLES – PYTHON

Python, or any code-based approach to analytics, whether it's bash scripts or an ancient dialect of COBOL, can make data processing easier. It can also lead to getting lost in code complexities and missing the bigger picture. By applying the Data Flow Map Framework, I can better understand what the code is trying to accomplish, troubleshoot issues more effectively, and most importantly, communicate the story clearly. Three modes – Source, Focus, and Build – with nine actions make all the difference in the world.

In the next chapter, I present API examples.

CHAPTER 7

Cloud API

As friends, we engage in conversation with words. Perhaps it's in person over coffee, discussing the weather, sports, and politics. Maybe it's a text or two hundred between family members trying to coordinate Dad's 80th birthday party. Or perhaps you're shouting into the world's winds, conversing publicly through social media.

Your digital confab is conducted between computers using dozens, if not hundreds, of Application Programming Interfaces (APIs). If you're buying coffee, the cash register is highly likely to utilize APIs to perform its function, so you can't escape these systems.

APIs serve as a digital contract between two computers. They specify that if you ask a question, make changes, delete, or add information in a specified manner, the partner computer will respond accordingly. Most APIs are protected by an authentication scheme that regulates access and limits capabilities.

Most computer languages provide methods for interacting with APIs, and some include entire libraries dedicated to simplifying API usage. It's that pervasive.

API conversations can be in text or binary format. In this example, I will use an API that is readily available to anyone with a terminal program. The return values will be in a JSON text format and can be read by an editor or a command-line file viewer.

CHAPTER 7   CLOUD API

I'll be querying weather data provided by the National Weather Service at the weather.gov domain. Weather data is observed at stations, and most large airports have automated equipment that generates the data and posts it (using APIs) to the National Weather Service's servers.

Throughout this API example, I'll be using five command-line tools:

1. Shell scripting on the command line allows for the automatic execution of commands. Your computer will have different approaches depending on make and model, but almost every non-mobile computer has a shell and ways of executing commands.

2. `curl` – Goes out to the web and pulls the data from weather.gov in JSON format and stores it in text files through a bash script.

3. `. jq` – Short for JSON Query. Processes the returned files to pull out useful information.

4. `Miller` – A Swiss army knife of a tool that makes it easy to convert between JSON and CSV, and also join data along the way.

5. `CSVlens` – A formatting tool for viewing CSV in a tidy tabular format. You'll see this in the outputs.

These tools are readily available and open source – your platform will have different installation methods. They aren't the only tools that accomplish this task; combining other tools can achieve the same results. These are the ones that best demonstrate the Data Flow Map and, honestly, the ones I understand the most.

My goal in this analysis is to get the weather information for a distribution of weather stations across the United States. While local weather data is readily available, I would like to take a more national look at the top hot or cold states and identify patterns there. In this analysis,

CHAPTER 7   CLOUD API

my final goal is to get the top five hot states. Once I've built a functional pipeline, I can then expand the analysis to a bigger picture of national weather, noting the highs and lows and other weather attributes.

I have a list of stations along with their corresponding states. I want a quick snapshot of the temperature and the station's state (Minnesota, Texas, etc.). I want to sort the results by temperature from high to low.

As with many command-line interface (CLI) jobs, I must work on this in phases. In a CLI sequence, you can pipe the results from one command to the next; I'll do that at the very end after I've built up the three major processes.

Each segment will include elements of the DFM throughout. For example, initially, I will use a CLI tool to access the API and pull data, but I will also do this for multiple locations.

I'll source data from the National Weather Service, a free API service available to anyone who needs the information. There are no limits to the data within reasonable boundaries, and you don't need to authenticate, which is a requirement for many APIs, primarily commercial ones. Getting your local weather is simply a matter of identifying a nearby large airport. They almost always have automated weather stations, and the NWS provides a free link to the data. You'll surely get it if you Google your airport, followed by the weather station name.

@ Land the Data
Source
o Get

Weather data is collected at stations in airports across the United States. I'll choose the first example from Minneapolis/Saint Paul Airport in Minnesota – KMSP. Obtaining the raw data is as simple as executing the following `curl` command in the command line, along with a destination file name, see Figure 7-1.

121

CHAPTER 7 CLOUD API

Listing 7-1. Curling Data Edges

```
curl https://api.weather.gov/stations/KMSP/observations/latest > ./KMSP.json
```

This returns over 150 lines of JSON data, starting with that shown in Figure 7-1.

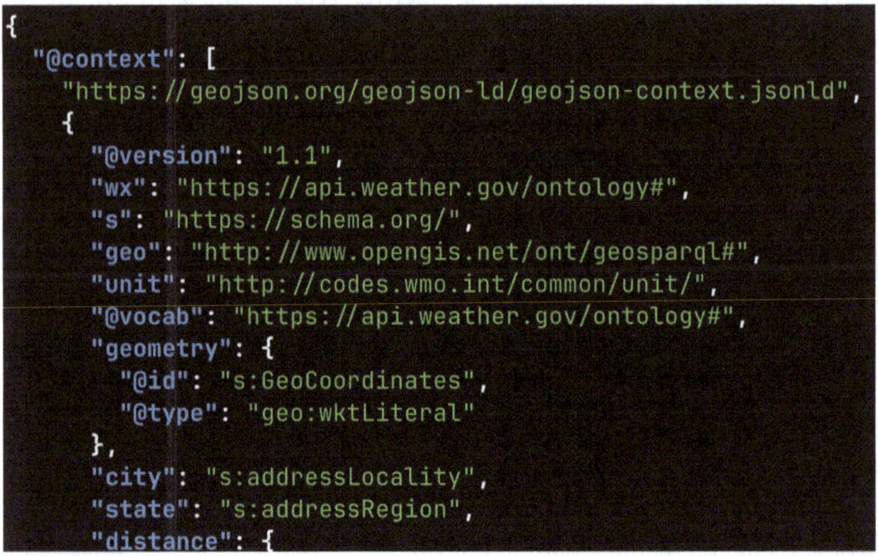

```
{
  "@context": [
    "https://geojson.org/geojson-ld/geojson-context.jsonld",
    {
      "@version": "1.1",
      "wx": "https://api.weather.gov/ontology#",
      "s": "https://schema.org/",
      "geo": "http://www.opengis.net/ont/geosparql#",
      "unit": "http://codes.wmo.int/common/unit/",
      "@vocab": "https://api.weather.gov/ontology#",
      "geometry": {
        "@id": "s:GeoCoordinates",
        "@type": "geo:wktLiteral"
      },
      "city": "s:addressLocality",
      "state": "s:addressRegion",
      "distance": {
```

Figure 7-1. *First Taste of the Data*

I can munge through this data using the jq tool by saving it to a text file named KMSP.json with the station's name. The "jq" tool is great for parsing, filtering, and transforming JSON data.

Since I want only the lines relevant to my work, not the entire JSON file, I'll focus on limiting the results.

CHAPTER 7 CLOUD API

@ Pull Multiple Stations

Now that I've landed at one station, I'd like to pull the data from multiple stations in one shot. For the purposes of this chapter, I've collected a list of stations across the nation that I thought would be useful for the analysis. It's subjective and mostly covers the range of the United States. The station list is kept in a stations.txt file that looks like Figure 7-2.

```
station_id    state_name    city_name                url
KATL          Georgia       Atlanta                  https://api.weather.gov/stations/KATL/observations/latest
KBGR          Maine         Bangor                   https://api.weather.gov/stations/KBGR/observations/latest
KCRP          Texas         Corpus Christi           https://api.weather.gov/stations/KCRP/observations/latest
KDFW          Texas         Dallas-Fort Worth        https://api.weather.gov/stations/KDFW/observations/latest
KGPI          Montana       Kalispell                https://api.weather.gov/stations/KGPI/observations/latest
KIAH          Texas         Houston                  https://api.weather.gov/stations/KIAH/observations/latest
KINL          Minnesota     International Falls      https://api.weather.gov/stations/KINL/observations/latest
KJFK          New York      New York                 https://api.weather.gov/stations/KJFK/observations/latest
KLAX          California    Los Angeles              https://api.weather.gov/stations/KLAX/observations/latest
KMCI          Missouri      Kansas City              https://api.weather.gov/stations/KMCI/observations/latest
KMIA          Florida       Miami                    https://api.weather.gov/stations/KMIA/observations/latest
KMSP          Minnesota     Minneapolis-Saint Paul   https://api.weather.gov/stations/KMSP/observations/latest
KORD          Illinois      Chicago                  https://api.weather.gov/stations/KORD/observations/latest
KPDX          Oregon        Portland                 https://api.weather.gov/stations/KPDX/observations/latest
KPHX          Arizona       Pheonix                  https://api.weather.gov/stations/KPHX/observations/latest
KSEA          Washington    Seattle                  https://api.weather.gov/stations/KSEA/observations/latest
KSFO          California    San Francisco            https://api.weather.gov/stations/KSFO/observations/latest
PANC          Alaska        Anchorage                https://api.weather.gov/stations/PANC/observations/latest
PHNL          Hawaii        Honolulu                 https://api.weather.gov/stations/PHNL/observations/latest
```

Figure 7-2. Station Number Nine

Source

X Mix

I will process each URL using a Bash script and send the results to a working folder.

Listing 7-2. Bash Script Processing Multiple Files

```
#!/bin/bash

# Read the stations.csv file line by line,
# skipping the header
```

CHAPTER 7   CLOUD API

```
while IFS=',' read -r station_id state_name city_name URL; do
    # Trim whitespace from the URL
    url=$(echo "$url" | xargs)

    # Fetch the data from the URL
    response=$(curl -s "$url")

    # Check if the response is valid JSON
    if echo "$response" | jq empty > /dev/null 2>&1; then
            # Save the response to a JSON file
            # named after the station_id
        echo "$response" > "data/${station_id}.json"
        echo "Saved data for station: $station_id"
    else
        echo "Failed to fetch or parse data for station:
        $station_id"
    fi
done < <(tail -n +2 stations.csv)
```

As a result, I should see a listing of JSON files like this in Figure 7-3.

CHAPTER 7   CLOUD API

```
.rw-r--r--@ 4.6k nick 18 Apr 21:06 {} KATL.json
.rw-r--r--@ 4.6k nick 18 Apr 21:06 {} KBGR.json
.rw-r--r--@ 5.0k nick 18 Apr 21:06 {} KCRP.json
.rw-r--r--@ 5.0k nick 18 Apr 21:06 {} KDFW.json
.rw-r--r--@ 4.6k nick 18 Apr 21:06 {} KGPI.json
.rw-r--r--@ 5.0k nick 18 Apr 21:06 {} KIAH.json
.rw-r--r--@ 4.6k nick 18 Apr 21:06 {} KINL.json
.rw-r--r--@ 5.0k nick 18 Apr 21:06 {} KJFK.json
.rw-r--r--@ 5.0k nick 18 Apr 21:06 {} KLAX.json
.rw-r--r--@ 4.8k nick 18 Apr 21:06 {} KMCI.json
.rw-r--r--@ 4.6k nick 18 Apr 21:06 {} KMIA.json
.rw-r--r--@ 5.0k nick 18 Apr 21:06 {} KMSP.json
.rw-r--r--@ 5.0k nick 18 Apr 21:06 {} KORD.json
.rw-r--r--@ 4.6k nick 18 Apr 21:06 {} KPDX.json
.rw-r--r--@ 4.8k nick 18 Apr 21:06 {} KPHX.json
.rw-r--r--@ 4.6k nick 18 Apr 21:06 {} KSEA.json
.rw-r--r--@ 4.6k nick 18 Apr 21:06 {} KSFO.json
.rw-r--r--@ 5.2k nick 18 Apr 21:06 {} PANC.json
.rw-r--r--@ 5.0k nick 18 Apr 21:06 {} PHNL.json
```

Figure 7-3. Listing of Files

Since the data is clean enough to move forward, I will not apply a "Fix" action at this time.

Focus

v Cut

I want to select just a few elements of the resulting JSON for this analysis, and jq makes that very easy.

- Station Identification – name, altitude
- Timestamp
- Temperatures – current, wind chill, heat index

- Humidity
- Wind data
- Visibility
- Text description of current weather

Since the JSON result contains much more than that, I will use the query part of jq to pull just the fields I want out of the complete JSON. To make it simpler and easier to maintain, I'm storing the query in a separate file called query.jq (Listing 7-2), and I'll ask jq to use that query while looking at the JSON file (Listing 7-3), and run the query in the command line (Listing 7-4).

Listing 7-3. jq Query Text

```
{
  Station_id: .station_id,
  elevation: .properties.elevation.value,
  station: .properties.station,
  timestamp: .properties.timestamp,
  relativeHumidity:. properties.relativeHumidity.value,
  windChill: .properties.windChill.value,
  heatIndex: .properties.heatIndex.value,
  windSpeed: .properties.windSpeed.value,
  windDirection: .properties.windDirection.value,
  temperature: .properties.temperature.value,
  visibility: .properties.visibility.value,
  textDescription: .properties.textDescription
}
```

Listing 7-4. Invoke jq with Query

```
jq -f query.jq ./KMSP.json
```

This gives a much more succinct and cleaner result; Listing 7-5.

Listing 7-5. Cleaning Up

```
{
  "station": "https://api.weather.gov/stations/KMSP",
  "elevation": 255,
  "timestamp": "2025-02-19T10:53:00+00:00",
  "relativeHumidity": 67.267750109568,
  "windChill": -31.771636240815557,
  "heatIndex": null,
  "windSpeed": 12.96,
  "windDirection": 310,
  "temperature": -22.8,
  "visibility": 16090,
  "textDescription": "Clear"
}
```

I will build this into a final result and ship it as a text file for the next go-around with Miller.

Build

Saving this into a result file is easy at this point.

= Ship

I want to process all JSON files simultaneously and drop them into a "combined.json" file. I'll adapt the jq code above to run through all the files; Listing 7-6.

CHAPTER 7   CLOUD API

Listing 7-6. Multiple Files into One

```
for file in ./data/*.json; do
    jq -f query.jq "$file"
done > ./clean_json/combined.json
```

Now all of the JSON files are concatenated into one output; Figure 7-4.

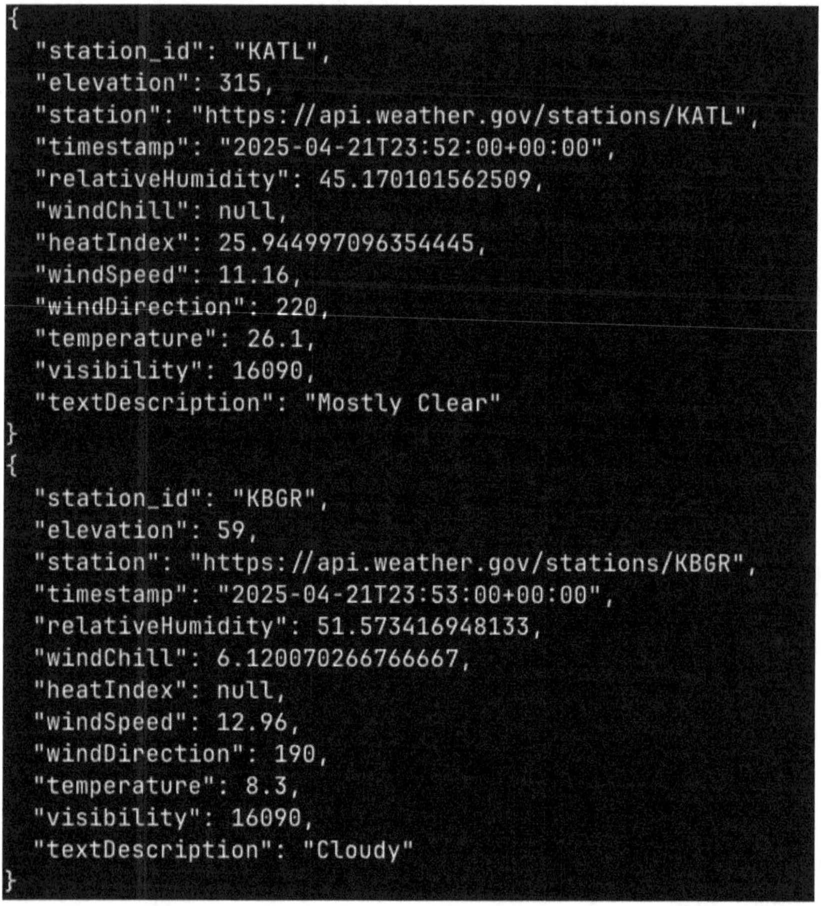

```
{
  "station_id": "KATL",
  "elevation": 315,
  "station": "https://api.weather.gov/stations/KATL",
  "timestamp": "2025-04-21T23:52:00+00:00",
  "relativeHumidity": 45.170101562509,
  "windChill": null,
  "heatIndex": 25.944997096354445,
  "windSpeed": 11.16,
  "windDirection": 220,
  "temperature": 26.1,
  "visibility": 16090,
  "textDescription": "Mostly Clear"
}
{
  "station_id": "KBGR",
  "elevation": 59,
  "station": "https://api.weather.gov/stations/KBGR",
  "timestamp": "2025-04-21T23:53:00+00:00",
  "relativeHumidity": 51.573416948133,
  "windChill": 6.120070266766667,
  "heatIndex": null,
  "windSpeed": 12.96,
  "windDirection": 190,
  "temperature": 8.3,
  "visibility": 16090,
  "textDescription": "Cloudy"
}
```

Figure 7-4. *Multiple Stations, One File*

CHAPTER 7 CLOUD API

@ Process JSON to CSV

I aim to have a simple CSV file with all the stations.

Source
O Get

I'll use the "combined.json" file from above.

X Fix

At this point, I can use the "jq" tool again to process the combined stations, as seen in Listing 7-7.

Listing 7-7. Process the JSON into CSV

```
jq -r '(input | keys) as $keys
| $keys | @csv, (inputs | [.[$keys[]]]
| @csv)' combined.json > weather_stations.csv
```

This generates the result in Figure 7-5.

elevation	heatIndex	relativeHumidity	station	station_id
13	24.63068991601	87.515784017394	https://api.weather.gov/stations/KCRP	KCRP
182	24.27533064050389	27.565854317177	https://api.weather.gov/stations/KDFW	KDFW
906		53.191916466365	https://api.weather.gov/stations/KGPI	KGPI
33	28.560328252597778	47.557922512366	https://api.weather.gov/stations/KIAH	KIAH
361		32.808572074644	https://api.weather.gov/stations/KINL	KINL
7		71.300446309395	https://api.weather.gov/stations/KJFK	KJFK
32		69.876502911497	https://api.weather.gov/stations/KLAX	KLAX
312		58.55887506803	https://api.weather.gov/stations/KMCI	KMCI
4	24.571937510630555	64.201862109253	https://api.weather.gov/stations/KMIA	KMIA
255		28.125875270292	https://api.weather.gov/stations/KMSP	KMSP
205		58.917436953004	https://api.weather.gov/stations/KORD	KORD
12		42.218135507507	https://api.weather.gov/stations/KPDX	KPDX
337			https://api.weather.gov/stations/KPHX	KPHX
137		40.073323923865	https://api.weather.gov/stations/KSEA	KSEA
5		60.799253674469	https://api.weather.gov/stations/KSFO	KSFO
44			https://api.weather.gov/stations/PANC	PANC
5	31.34803133702889	46.385645157686	https://api.weather.gov/stations/PHNL	PHNL

Figure 7-5. jq Simple Output

CHAPTER 7   CLOUD API

+ Mix

Now I want to return the original station location to the mix.

Several tools will do this on the fly – I will use Miller (mlr) to do this (Listing 7-8).

Listing 7-8. Join the Weather csv Back to the Stations csv

```
mlr --csv join -j station_id -f stations.csv then \
  cut -x -f url then \
  weather_stations.csv > combined_weather_data.csv
```

This joins the two files on the station_id column and drops a duplicate URL field (Figure 7-6).

station_id	state_name	city_name	elevation	heatIndex	relativeHumidity
KCRP	Texas	Corpus Christi	13	24.63068991601	87.515784017394
KDFW	Texas	Dallas-Fort Worth	182	24.27533064050389	27.565854317177
KGPI	Montana	Kalispell	906		53.191916466365
KIAH	Texas	Houston	33	28.560328252597778	47.557922512366
KINL	Minnesota	International Falls	361		32.808572074644
KJFK	New York	New York	7		71.300446309395
KLAX	California	Los Angeles	32		69.876502911497
KMCI	Missouri	Kansas City	312		58.55887506803
KMIA	Florida	Miami	4	24.571937510630555	64.201862109253
KMSP	Minnesota	Minneapolis-Saint Paul	255		28.125875270292
KORD	Illinois	Chicago	205		58.917436953004
KPDX	Oregon	Portland	12		42.218135507507
KPHX	Arizona	Pheonix	337		
KSEA	Washington	Seattle	137		40.073323923865
KSFO	California	San Francisco	5		60.799253674469
PANC	Alaska	Anchorage	44		
PHNL	Hawaii	Honolulu	5	31.34803133702889	46.385645157686

Figure 7-6. *Combined Station and Data*

@ Top Five States

We're almost at the end. I want a simple file with the station_id, the state, and the temperature, which should be Fahrenheit. This can all be done in Miller.

CHAPTER 7　CLOUD API

Source

O Get

Use the combined_weather_data.csv

X Fix

Using inline Miller functions, convert the Celsius temperature to Fahrenheit; Listing 7-9.

Listing 7-9. *Specific Code to Convert C to F*

```
mlr --csv put '$temp_F = ($temperature * 9/5) + 32;
$temp_F_round = round($temp_F)'
combined_weather_data.csv > fix_combined_weather.csv
```

This will add a new column named temp_F_round that we can use for easier reporting; see Figure 7-7.

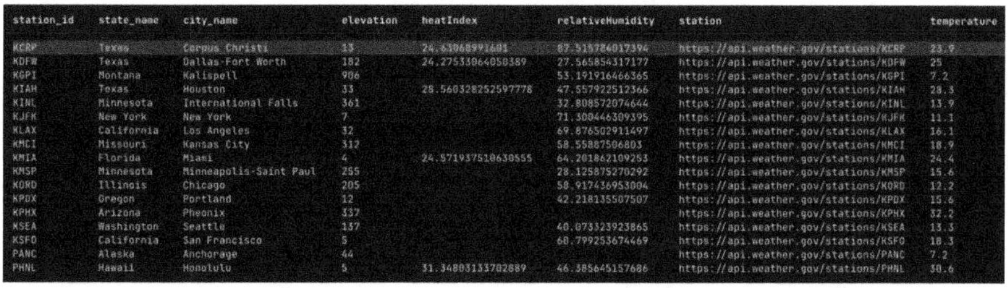

Figure 7-7. *New Round Column*

Focus

V Cut

Let's pick the station_id, the state name, and the final temperature, and put the results in a working file, as seen in Listing 7-10.

131

CHAPTER 7 CLOUD API

Listing 7-10. Cut Columns

```
mlr --csv cut
-f state_name,temp_F_round ./ fix_combined_weather.csv > cut_
combined_weather.csv
```

This results in Figure 7-8.

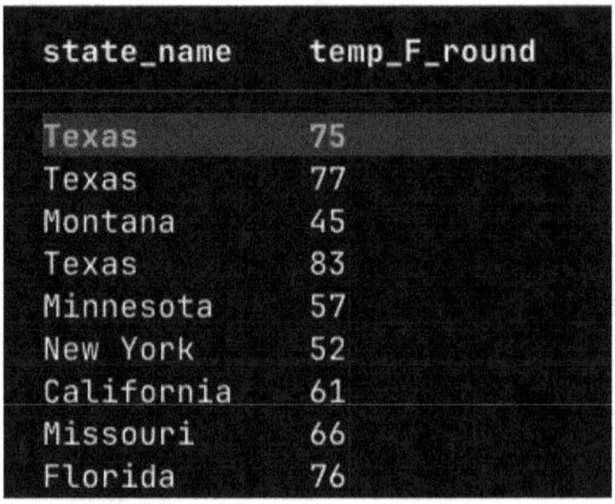

state_name	temp_F_round
Texas	75
Texas	77
Montana	45
Texas	83
Minnesota	57
New York	52
California	61
Missouri	66
Florida	76

Figure 7-8. Cut Columns Result

Build

There is some good information here, but I would like to break it down to the state level, finding the maximum value for each state. For grouping categories and sizing values, Miller wants to do this in one step using "stats1" and the "-g" flag, as seen in Listing 7-11.

] Box and # Size

Listing 7-11. Group By State and Size Temp

```
mlr --csv stats1 -a max -f temp_F_round -g state_name ./sort_
combined_weather.csv > box_combined_weather.csv
```

This results in Figure 7-9.

```
state_name            temp_F_round_max

Alaska                45
Arizona               90
California            65
Florida               76
Hawaii                87
Illinois              54
Minnesota             60
Missouri              66
Montana               45
New York              52
Oregon                60
Texas                 83
Washington            56
```

Figure 7-9. High Temp for Each State

I now have only 13 entries, one for each state.

= Sort

The goal of this analysis is to get the top five temperatures by state, so I'll apply a sort by temperature, descending values, as seen in Listing 7-12.

Listing 7-12. Final Result

```
mlr --csv sort -nr temp_F_round_max ./box_combined_weather.csv |
head -n 6 > sort_combined_weather.csv
```

This results in the output in Figure 7-10.

CHAPTER 7   CLOUD API

```
state_name          temp_F_round_max

Arizona             90
Hawaii              87
Texas               83
Florida             76
Missouri            66
California          65
Minnesota           60
Oregon              60
Washington          56
Illinois            54
New York            52
Alaska              45
Montana             45
```

Figure 7-10. *Getting Closer*

= Ship

Finally, at the end of the day, I just want to have the top five entries. I'm going to jump out of Miller and go back to "head" on the command line. Note that I'll have to grab six lines to capture the header row, as seen in Listing 7-13.

Listing 7-13. *Using Head to Limit, with the result see in Image 7-11.*

head -n 6 ./sort_combined_weather.csv > ship_combined_weather.csv

CHAPTER 7 CLOUD API

```
state_name         temp_F_round_max

Arizona            90
Hawaii             87
Texas              83
Florida            76
Missouri           66
```

Figure 7-11. Final Result

Summary

You can apply the same logic and approach of the Data Flow Map to a process entirely via the command line. The steps could be concatenated at this stage into a single shell script that pulls the data, processes it, and generates the final output.

In this case, I used the following stages to go from getting the data off the internet to the final result:

1. **Land the Data:** Pull one station from the internet locally

2. **Pull Multiple Stations:** Pull a range of stations from the internet

3. **Process JSON to CSV:** Convert the format

4. **Top Five States:** Roll up the values by state and pick top five

Each of these four stages used the Source, Focus, and Build modes with the nine actions to process the data.

CHAPTER 7   CLOUD API

Iterating through the process of using command-line tools can be frustrating due to errors and syntax mistakes. Utilizing the Data Flow Map can segment the work more clearly, making testing and learning from individual examples easier before attempting the entire process in one fragile step. Since the intermediary stages are built on JSON and then converted to CSV, numerous tools can be applied at almost any time.

In the next chapter, I present platforms.

CHAPTER 8

Platforms

The concept of platform-based analytics has been around for decades. Imagine an application running on a local computer that will give you drag-and-drop access to analytics. Now, supercharge it with dazzling graphics that make any beginner analyst look like a rock star on their first gig at the local amusement park.

As most desktop applications have moved to the web, so have platform-based analytics. There are many contenders – some old, some new, and some still trying hard to make a difference.

A few of the bigger ones are:

- **Tableau**: One of the glossiest, they have mastered the fine art of clean graphics and drag-and-drop.

- **PowerBI**: Microsoft's entry is a solid contender and is continually updated and improved.

- **Power Query**: Built into every Excel app in the land, it's the platformer tool you didn't know you had at your fingertips.

- **Looker**: Google's entry in a world of glamorous charts and dashboards.

- **Superset**: By Apache, as one of the few contending open-source projects.

CHAPTER 8   PLATFORMS

Most of these tools excel at presenting data, and all are less adept at initial sourcing, particularly merging two or more sources. This chapter will utilize Superset and its built-in SQL interface for data sourcing. I'll explore the drag-and-drop nature of platform analytics and demonstrate some of the glossiness that they provide.

I'll use some straightforward geographic data for Minnesota cities with the word "lake" in their name for sample data. Minnesota's slogan is "Land of 10,000 lakes," and while that's a slight understatement, depending on your definition of a lake, it's a part of the culture. I am interested in several themes from the data:

- Lake towns, in general, tend to be sleepy weekend tourist communities. Is there a tie-in between the current population and the distance to the metro area? In other words, are Lake towns turning into bedroom communities more interested in soccer than fishing?

- Have Lake towns grown or shrunk in general over the last several decades?

Platform tools are excellent for this type of "what if" analysis, and the speed at which you can try out theories and get traction can be intoxicating.

Source

Most data sourcing work will be in the provided "SQL Lab" tab. This tab lets users get, fix, and mix the data to their hearts' content (Figure 8-1).

CHAPTER 8   PLATFORMS

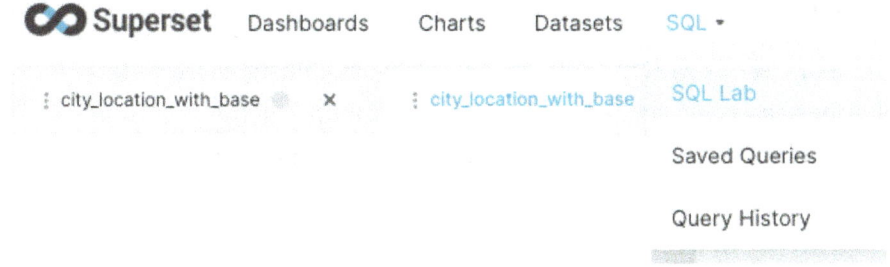

Figure 8-1. Navigate to the SQL Lab Tab

This instance of Superset runs with a Postgres database that contains the sample city data.

0 Get

The original source for this list of lake-based cities was Wikipedia, and the decade census data for 1970 forward was pulled manually in an initial CSV file called "city_population.csv". When I was gathering the data, I wasn't sure how the information would be put together, so I labeled the decade as "attribute" and the population as "value" – something I'll have to fix shortly.

Table 8-1. city_population

City	Attribute	Value
Lakeville	1970	7556
Lakeville	1980	14790
Lakeville	1990	24854

As you can see in Figure 8-1, Lakeville experienced significant growth in the first few decades and continued to skyrocket.

139

CHAPTER 8   PLATFORMS

X Fix

There are issues with the column names; worse, Postgres isn't a fan of upper-case field names. In the SQL tool, I'll update the field names accordingly and set it up for the mix coming up; in Figure 8-2.

Figure 8-2. First Query to Fix Column Names

In my initial survey of the data, I want to identify any larger patterns in the town's population. While the data isn't extensive, it's more than I can keep in mind, so I'll add a categorical value to bucket the town sizes later. At this point, I'm not sure if I'll use this in the final result, but it will be another angle to explore.

I will add a categorical "CASE WHEN" (Figure 8-3).

CHAPTER 8   PLATFORMS

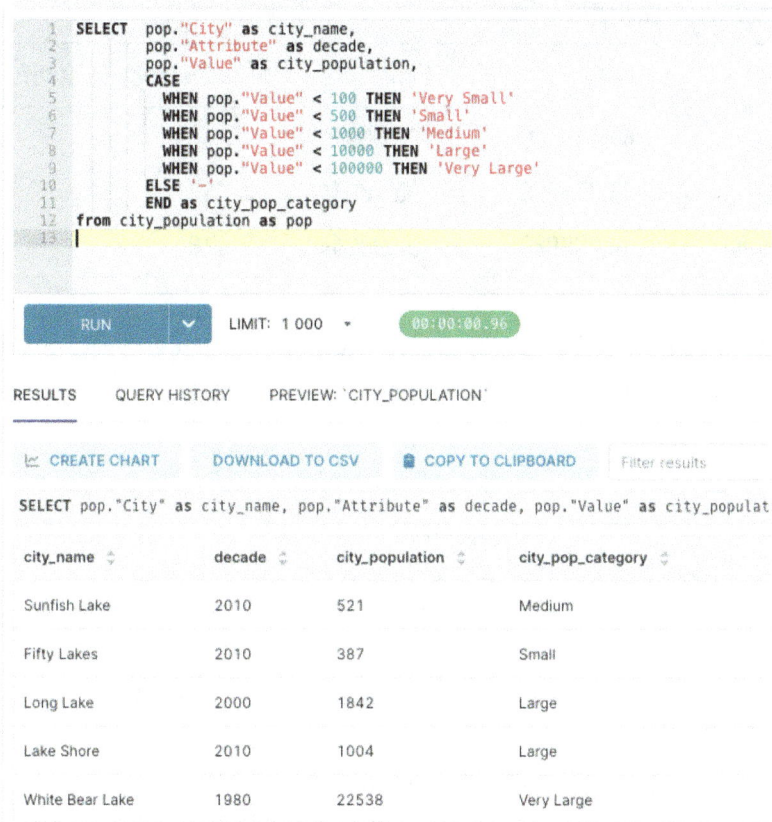

Figure 8-3. Population with Categories

X Mix

I want to add context regarding the distance to the nearest central metropolitan area. To help get there, I compiled a table of the Lake Cities, including their latitude and longitude, as well as the distance to the metro area, using mapping resources. This table will be called city_location_distance.

CHAPTER 8 PLATFORMS

Table 8-2. Example Cities

City	Lat	Lon	Distance to Metro
Lake City	44.445556	-96.289167	163
Lakefield	43.678056	-95.169444	131
Lake Elmo	44.998889	-92.909444	13

Using SQL, I'll join the two tables, merging the location, distance, decade, and population. See Listing 8-1.

Listing 8-1. Mix

```
SELECT  pop."City" as city_name,
        pop."Attribute" as decade,
        pop."Value" as city_population,
        CASE
          WHEN pop."Value" < 100 THEN 'Very Small'
          WHEN pop."Value" < 500 THEN 'Small'
          WHEN pop."Value" < 1000 THEN 'Medium'
          WHEN pop."Value" < 10000 THEN 'Large'
          WHEN pop."Value" < 100000 THEN 'Very Large'
        ELSE '-'
        END as city_pop_category,
        dst.Lat as latitude,
        dst.Lon as longitude,
        dst.Distance_To_Metro as distance
from city_population as pop
JOIN city_location_distance as dst
ON pop.City = dst.City
```

This results in Figure 8-4.

CHAPTER 8   PLATFORMS

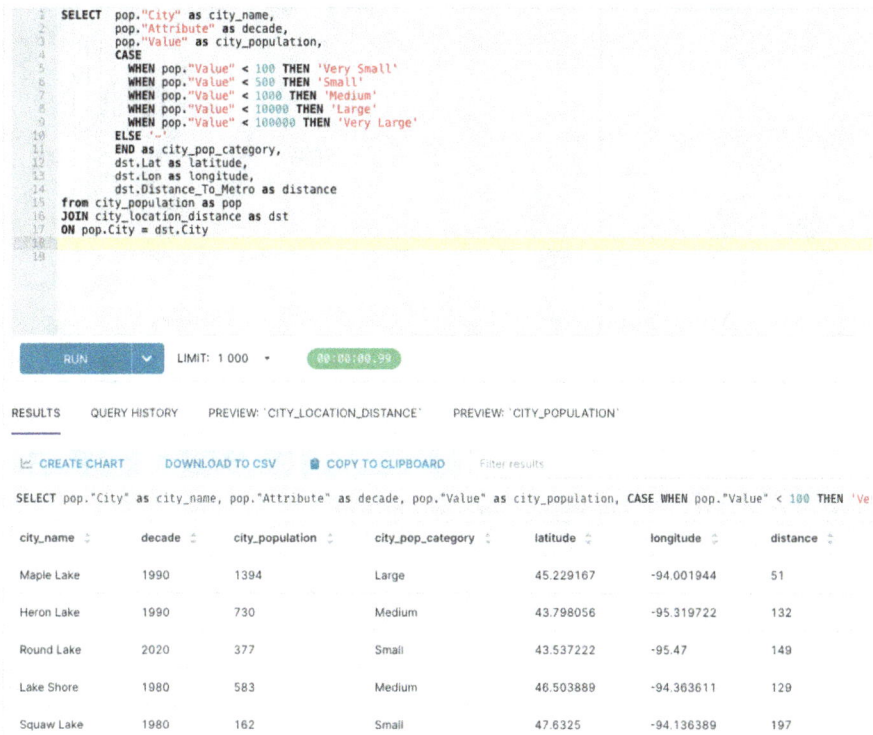

Figure 8-4. Mixing in the City Location Data

At this point, you might be wondering where the "platform" piece comes into the conversation. The above code could be run from virtually any SQL client. Ah, but this is where life gets interesting.

I'll save the query and then the dataset. This will make it available for graphing in the Superset Chart tool (Figure 8-5).

CHAPTER 8   PLATFORMS

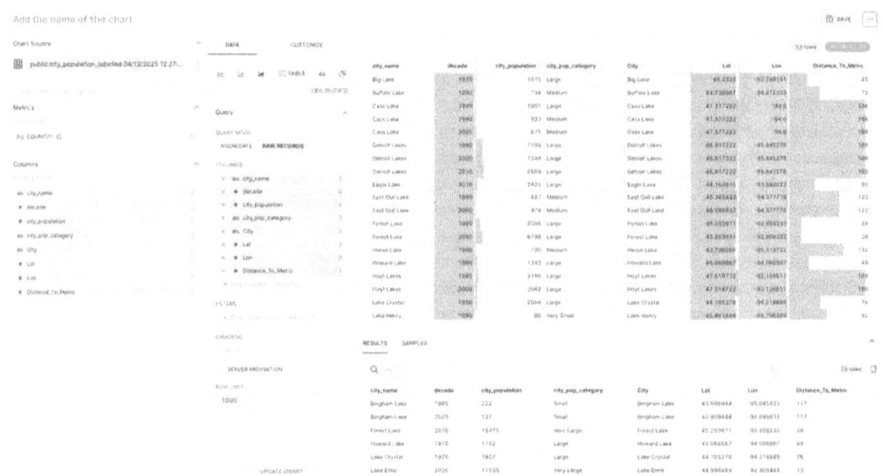

Figure 8-5. *Now We're Getting Somewhere*

The data has been sourced, fixed, and mixed. Let's start with a basic chart.

Focus

Most platform tools like this have a drag-and-drop method for importing data into a specific chart. For this first walkthrough, I will use a plain bar chart.

V Cut

Let's look at the city and population, and drop all the other extra stuff for now. The "Columns" box makes it easy to drop and add values (Figure 8-6).

CHAPTER 8 PLATFORMS

Figure 8-6. Getting Picky About Columns

If I limit it to the City and its population, I'll get a straightforward bar chart, as shown in Figure 8-7.

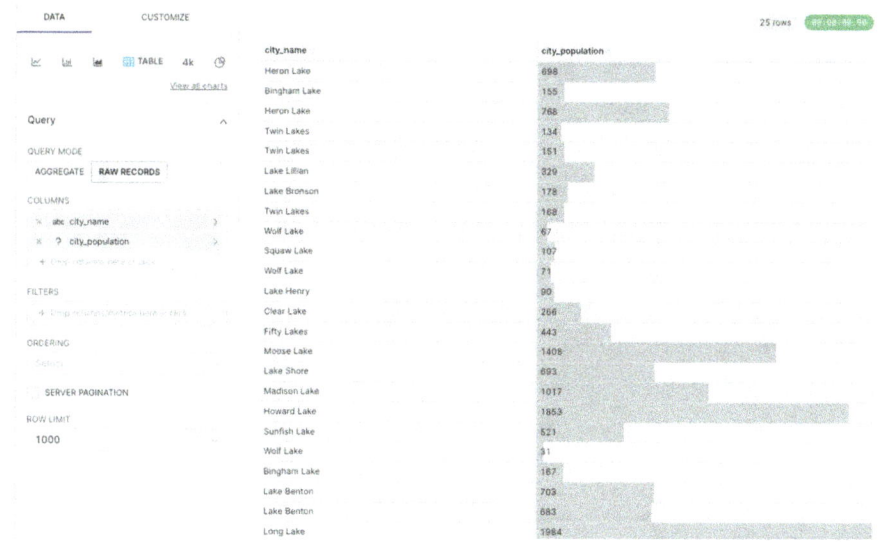

Figure 8-7. Getting More Focused

145

CHAPTER 8   PLATFORMS

> Slice

I don't want to have each decade – I'll stick with rows where it's the latest. I can drag and drop the decade field to the "FILTERS" box; Figure 8-8.

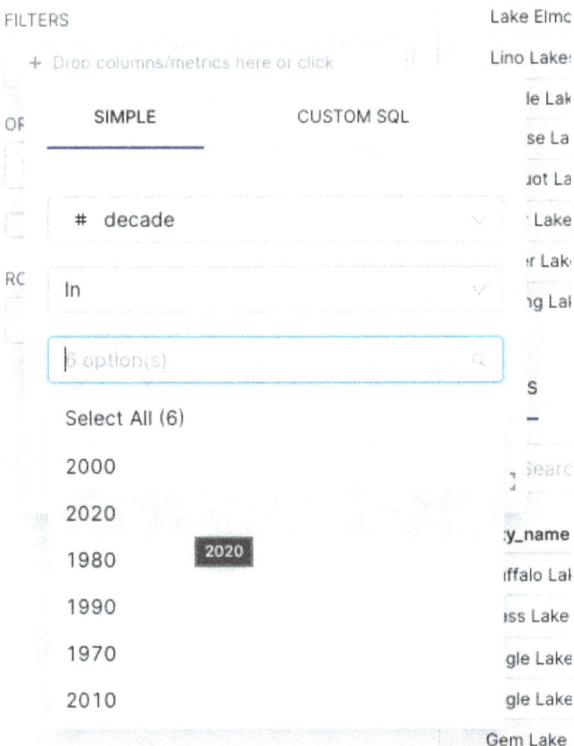

Figure 8-8. *Simple Filters*

CHAPTER 8   PLATFORMS

Much better, but it needs a little more work; Figure 8-9.

Figure 8-9. *Just 2020 Data*

\ Sort

Now that I've got the latest population, I want to sort top to bottom on population. I'll use the "ORDERING" drop-down and pick the population (desc) option;

With the results seen in Figure 8-10.

CHAPTER 8   PLATFORMS

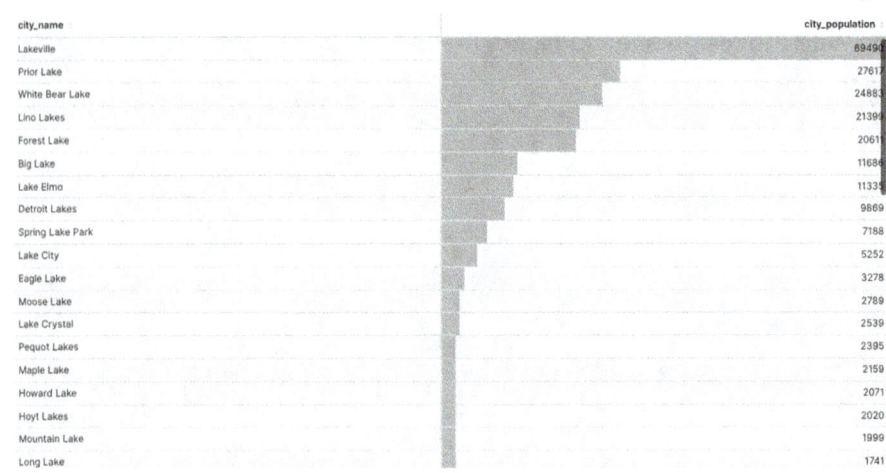

Figure 8-10. *Picking the Order*

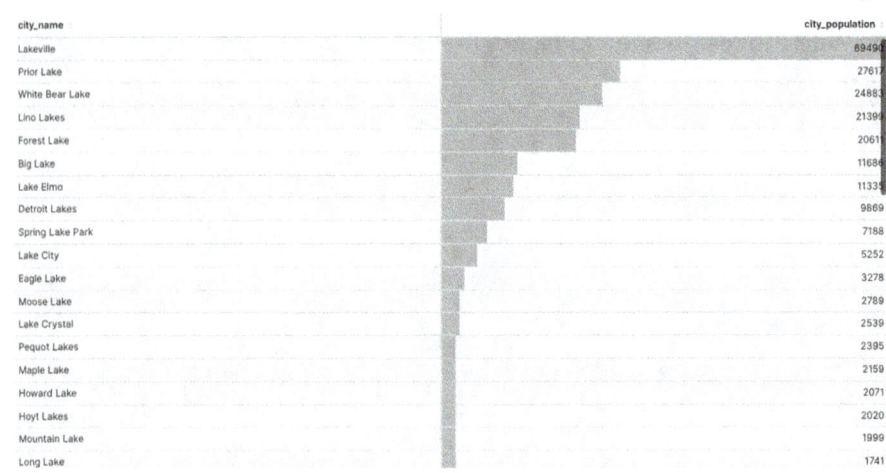

Figure 8-11. *Sorted by Descending Population*

CHAPTER 8   PLATFORMS

In the space of a few minutes, I can transform raw data into really serviceable charts. More importantly, I can pivot as I go through and learn more about the raw data. For instance, at a glance, I can tell that Lakeville is an outsider with nearly 70,000 residents.

Build

Now that I know how the populations play out, I'd like to see if there's a relationship between the city's population growth or decline and its distance to the metro area.

Superset provides an out-of-the-box Scatterplot that makes it simple to compare values.

] Box

I will just group by the cities' names for simplicity's sake.

Size

Since I will be looking at the growth (or reduction) in a city's size, I will have to compare the population in 1980 and 2020. This forty-year period should show the difference, and I want to demonstrate whether distance to the metro area matters.

I'm fixing my query to help create Figure 8-12 calculations.

CHAPTER 8   PLATFORMS

```
(SELECT *
 FROM fixed AS fxd
 JOIN public.city_location_distance AS dst ON fxd.city_name = dst."City"),
    pop_80 AS
(SELECT city_name,
        city_pop_category,
        "Distance_To_Metro",
        city_population AS city_pop_1980
 FROM mixed
 WHERE decade = 1980),
    pop_20 AS
(SELECT city_name,
        city_pop_category,
        "Distance_To_Metro",
        city_population AS city_pop_2020
 FROM mixed
 WHERE decade = 2020)
SELECT pop_80.city_name,
       pop_80."Distance_To_Metro" AS distance_to_metro,
       city_pop_1980,
       city_pop_2020,
       ((city_pop_2020::DOUBLE PRECISION - city_pop_1980::DOUBLE PRECISION)/(city_pop_1980::DOUBLE PRECISION)) * 100 AS city_pop_change_pct
FROM pop_80
JOIN pop_20 ON pop_80.city_name = pop_20.city_name
```

Figure 8-12.   Calculate the Percent Change

= Ship

The final result of the tool, provided at no cost, is shown in Figure 8-13.

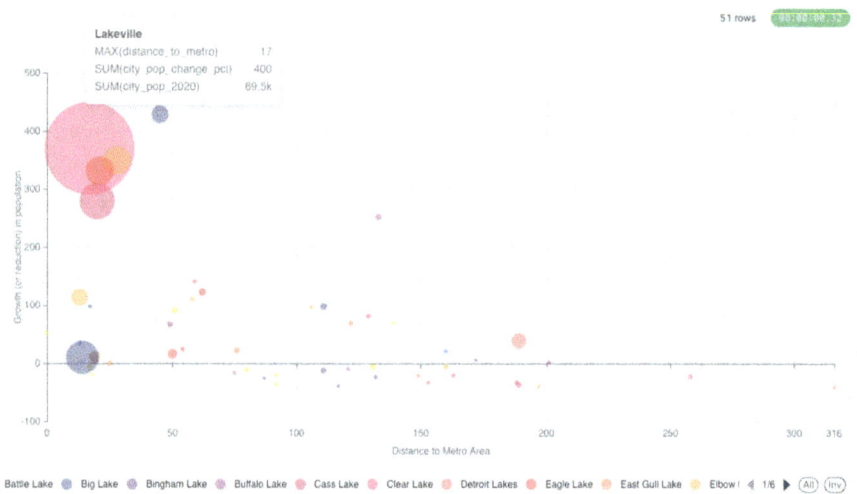

Figure 8-13.   Scatterplot of Scattered Thoughts

In this table, the bubble size indicates city population, and by hovering over them, I can see what city changed. For example, I can see that Lakeville is 400% bigger than it was in 1980. It's also an outlier. Just up and to the right is Big Lake, which has also grown significantly but not to the level of Lakeville.

150

CHAPTER 8 PLATFORMS

A different take on this will help answer the basic question at the beginning: Does distance from the metro correlate to growth and population? I'll use a scatterplot for this in Figure 8-14.

Figure 8-14. *Another Look*

Both population (Y-axis) and growth (bubble size) seem to indicate that the closer you are to town, the better chances you have to succeed.

The joy of these platform tools is that it makes it fairly easy to think about the data differently. For example, if I want to go back to the original question of "Have Lake towns grown or shrunk since 1980?", I can create a simple bar graph; see Figure 8-15.

151

CHAPTER 8   PLATFORMS

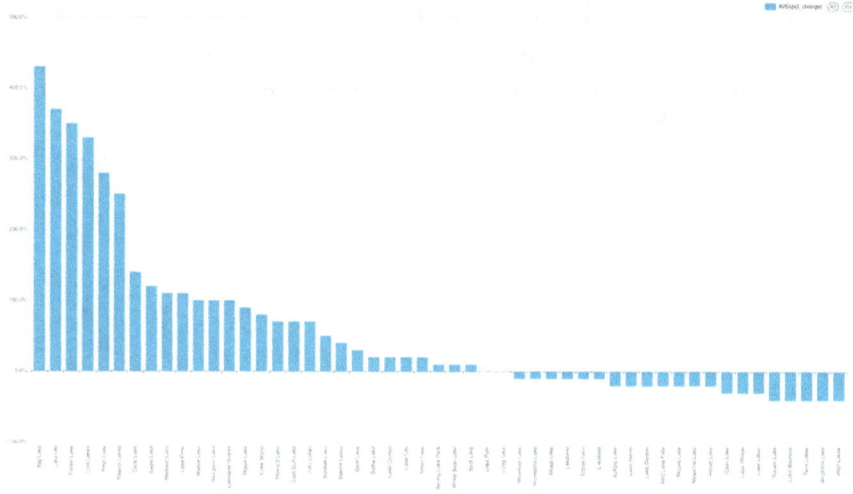

Figure 8-15.   Percent Change by Town

Most platform tools like this offer some dashboarding capabilities, allowing you to add multiple charts to a single pane. I've done this with several example charts showing how you can create chaos without trying; see Figure 8-16.

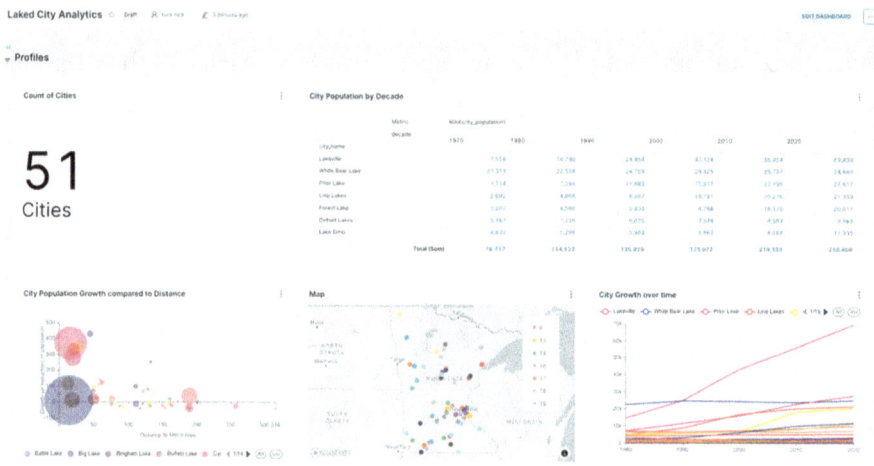

Figure 8-16.   Don't Do This, Even If You Can

Summary

Big picture, here's what I did in this chapter, and most of it took seconds to perform.

Motion	Symbol	Action	Goal
Source	O	Get	Pulled in the city location data
	X	Fix	Corrected labeling and categorized by size, calculated growth over the decades
	+	Mix	Mixed in distance data
Focus	V	Cut	Just city and population
	>	Slice	Just data from 2000
	\	Sort	Ordered by population, descending
Build	]	Box	Group by City and Category
	#	Size	Calculated growth versus distance along with bubble size on population
	=	Ship	Built scatterplot visual

I was able to iterate quickly when I got to a dead end and try out different ways of thinking about the data.

Platform tools are often costly to buy and require a lot of resources. The more design power you provide your end users, the faster they can create innovative solutions. However, there are several warnings most vendors won't mention.

- **The Lost Employee Problem**: Mary worked on this chart last year and now works for the competitor. How did she filter the data? I'm not sure, and platform charts are not always easy to understand in their implementation.

- **Competing Numbers**: With the ability to quickly create multiple charts and dashboards from the same base data, two or more users can come to different conclusions just because filters or groups are applied slightly differently. Where you have two sets of numbers, you now have no sets of numbers.

- **Performance Issues**: At first, the database and presentation layers will keep up with the end user demand – there will only be a few beta users. However, once you roll it out to the entire company, you must use heavy hardware and higher licensing fees to make a difference.

- **Data Preparation Tools Are Non-existent**: You will almost always have to have your data completely lined up, and the mental shift between a glossy graphic interface and old-school SQL can be jarring.

On the plus side:

- **Clean Charts Drive Behavior**: By default, they almost always have great looks – clean, simple, spare, and publisher-worthy.

- **Speed to Delivery**: Drag and drop is a valid metaphor; experimenting quickly with different analytics in different directions can be amazing.

- **Adoption Is Fast**: Onboarding a new user is often a matter of explaining the basics and then letting them learn on the job.

Platform tools are powerful, fast, and great-looking. Pair one with a pipeline tool like Knime or Alteryx, and you can conquer the world. Just don't try to scale it to the entire organization.

As with all of the other ways I can process data, it's easy to get lost in long rabbit holes exploring patterns. The final result can be glossy and easy on the eyes, but supporting the analysis with a clear summary on the Source, Focus choices, and final Builds can get lost and slow down creativity. The Data Flow Map Framework can help with this!

In the next chapter, I present pipelines.

CHAPTER 9

Examples – Pipelines

Pipelines

Much of what we do as analysts, engineers, and scientists resembles leaky plumbing. We pipe data from one location, prepare it in various ways, merge it with other sources, and ultimately fill the bathtub with soapy content. Hopefully, we're either having fun or at least getting clean.

The plumbing metaphor is old, and a few tool vendors make it incredibly easy to connect the pipes, attaching various nodes in remarkable ways. While the implementation is usually straightforward and quick, the actual result can resemble a bowl of spaghetti more than functional plumbing. That's okay because we can apply the same Data Flow Mapping techniques to clarify what's happening when and where, even further.

Two of the biggest pipeline-style tools in the market are KNIME and Alteryx. Both offer a visual node and connector-style method for processing data, a wide variety of tools for managing data, and online services useful for corporations needing to retain and share content. Other data-related tools have also provided more limited versions of the same concept. Typically, essential tools for managing data are offered for free, while bespoke, specialized tools are developed for more niche needs and made available in the community or for purchase.

CHAPTER 9   EXAMPLES – PIPELINES

In this example, I will be working with KNIME and solely use the off-the-shelf tools to source, focus, and build results from a relatively messy data set describing tenancy at the Mall of America. While the Mall of America is one of the largest malls in the world, it's not immune to the problem afflicting every mall: retaining tenant stores. Retail stores suffer the whims of time and fashion – what was once astonishing 30 years ago is now considered boomer cringe. Customers don't want that.

In 1992, the Mall of America was built with over 500 addressable store locations, over a hundred temporary cart spaces, four major department store anchors, and one theme park, all combined within over 5.5 million square feet. Over 30 years, the mall added two hotels, lost two anchor stores, survived COVID-19, remodeled several sections, and shifted its focus from a shopper's paradise to a tourist's selfie haven.

Our sample data captured the Mall's store listings at four points: 1994, 2007, 2017, and 2024. Since the data comes from publicly available sources, its quality and format have changed dramatically over the past thirty years, evolving from PDF to three completely different online formats. In a way, how the Mall of America tells the story of store locations also tracks technology changes. This data will reflect those changes from scanned paper to different ways the web has worked over the decades.

From this analysis, I would like to know how many stores that opened in 1994 still operate in the exact location under the same name. In a large mall like this, it's not uncommon for stores to relocate as their popularity fluctuates and owners' preferences shift. For example, Ragstock, a clothing recycler company, has occupied four locations since 1994. Since the data capture points are roughly every decade, the stores may have moved even more frequently. Additionally, I'd like to know if there's a relationship between the general location of the store and its ability to survive over the decades. The mall has four compass directions, each with a distinctly different style and vibe. Does that matter?

CHAPTER 9   EXAMPLES – PIPELINES

Source

The data comes in four text files, each with slightly different information, column names, and order, and contains extraneous information.

0 Get

Pulling the data in from CSV files is the easy part, as in Figure 9-1.

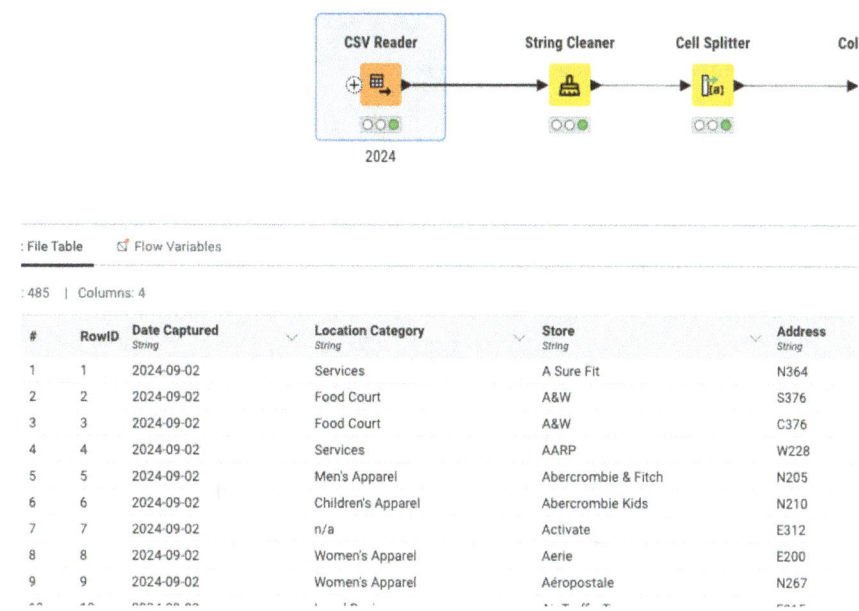

Figure 9-1. Load a CSV File

The nice thing about pipeline tools is that they make the data tangible. I can preview the data at almost every step and see if I'm heading in the right direction. As I apply fixes, joins, filters, and so on, the sample data at the bottom of the page will update with the currently selected mode. I can always click on a prior node to see how the data flows up to this point.

CHAPTER 9   EXAMPLES – PIPELINES

Remember that I'm working with a minimal data set, and your data needs will likely be more comprehensive. This is where sampling data can be invaluable.

I've created a CSV reader for each source and named it accordingly; Figure 9-2.

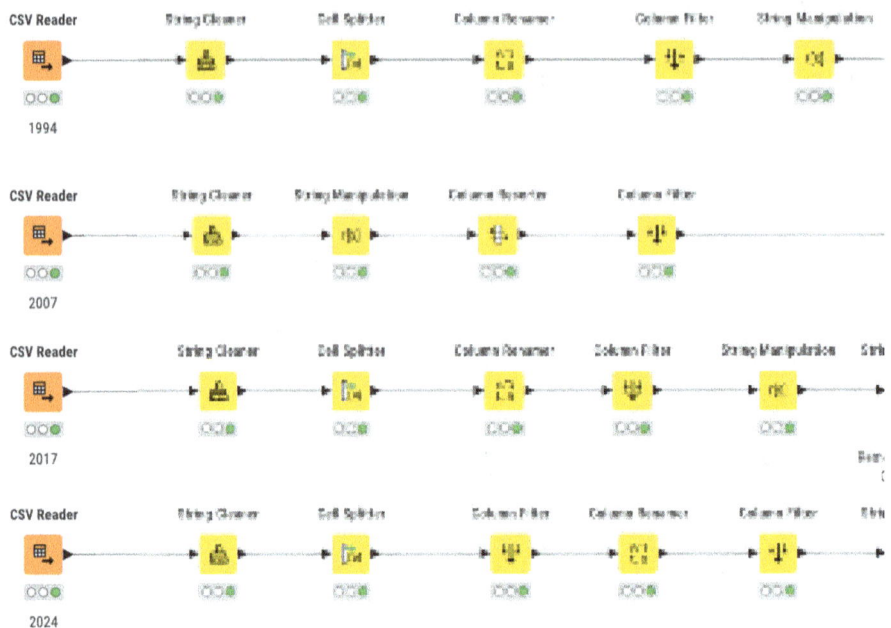

Figure 9-2. *Get Raw CSV File*

I can preview the data quickly by clicking on a node and executing. This is especially helpful for situations where the data doesn't line up correctly, has different column names, or is just extra. See Figure 9-3.

160

CHAPTER 9   EXAMPLES – PIPELINES

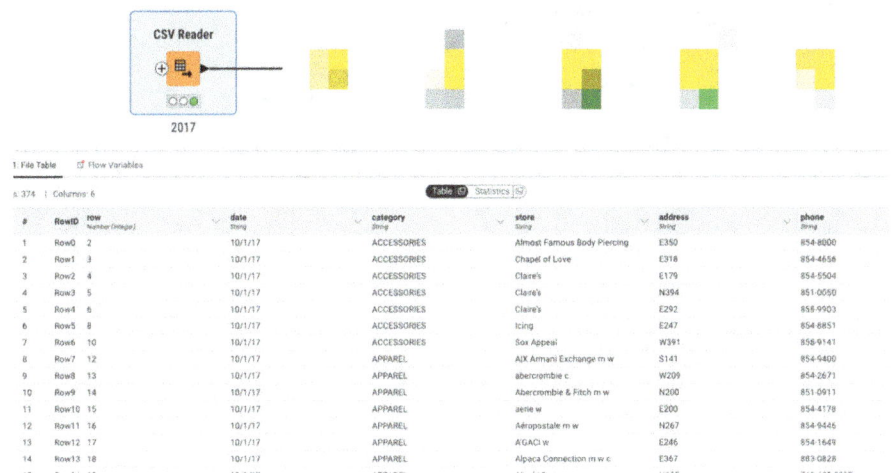

Figure 9-3. Previewing Data – Knime's Power Feature

+ Mix

One goal is to categorize the addresses by direction and floor. A reference table of all addresses is provided in a separate CSV file. This data looks like Figure 9-4.

#	RowID	Original_Address	Direction	Direction_Full	Floor	Room
1	Row0	E114	E	East	1	14
2	Row1	N394	N	North	3	94
3	Row2	E262	E	East	2	62
4	Row3	E168	E	East	1	68
5	Row4	E314	E	East	3	14
6	Row5	N365	N	North	3	65
7	Row6	W391	W	West	3	91
8	Row7	N294	N	North	2	94
9	Row8	W108	W	West	1	8
10	Row9	E266	E	East	2	66
11	Row10	W360	W	West	3	60
12	Row11	E275	E	East	2	75
13	Row12	W209	W	West	2	9
14	Row13	W144	W	West	1	44
15	Row14	N276	N	North	2	76

Figure 9-4. Sample of Mix

161

CHAPTER 9   EXAMPLES – PIPELINES

This is joined, as in Figure 9-5.

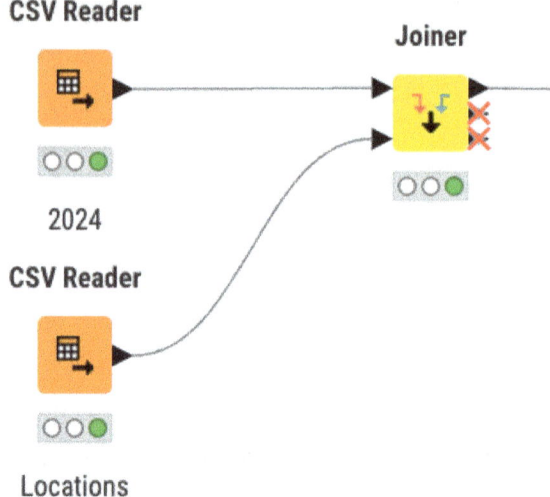

Figure 9-5. Join Node

Joins are configured in the settings dialog, accessible by double-clicking on it (see Figure 9-6).

CHAPTER 9   EXAMPLES – PIPELINES

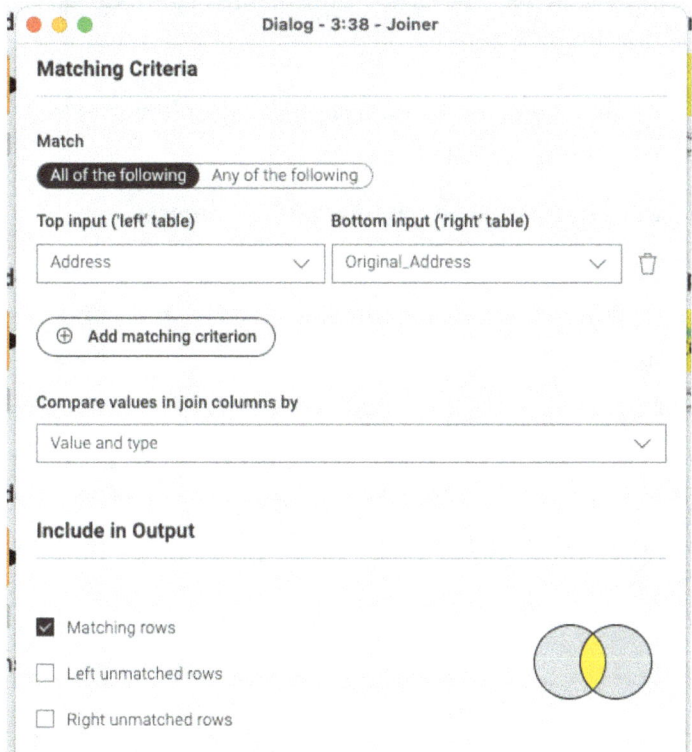

Figure 9-6. *A Typical KNIME Dialog Box – Showing Matching Criteria*

X Fix

One of the challenges of working with similar data from different sources – in this case, data from various decades – is maintaining consistency and ensuring availability. Column names change, the amount of detail included, and the order of columns are all issues that arise when combining data. Over the decades, it's entirely likely that at least four different primary authors were lining up the data.

CHAPTER 9　EXAMPLES – PIPELINES

To help drive consistency, several repairs need to be made to the data, including:

- The actual date of the data capture doesn't matter for this exercise – what's the year?

- Columns need to be renamed and re-ordered.

- The extra text needs to be removed. Different decades had different ways of identifying types of stores and locations.

These fixes can be made with a series of KNIME nodes that allow you to correct the data. For example, the date field in each file has a slightly different layout. I want to capture the year as the "year" field and correct junk data in the address. This can be done in a series of nodes, as Figure 9-7 shows.

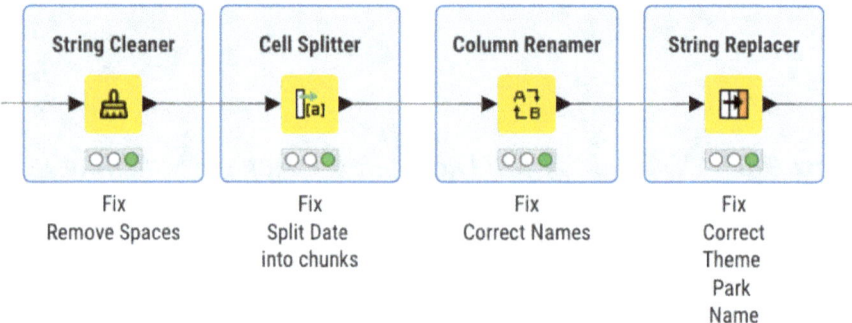

Figure 9-7. Chaining Fixes to the Data

An example of the fixes taking place, from the beginning to the end of a series of nodes, the data will start as Figure 9-8 and end as Figure 9-9.

CHAPTER 9   EXAMPLES – PIPELINES

Figure 9-8. First Fixes

Figure 9-9. Ready for Making Choices

Focus

While the data sets aren't large, they are messy and must be sorted to find the correct data.

V Cut

For the particular year I'm focusing on, there's a lot of extra data in the flow. It's time to narrow it down to just a few key fields in Figure 9-10.

CHAPTER 9   EXAMPLES – PIPELINES

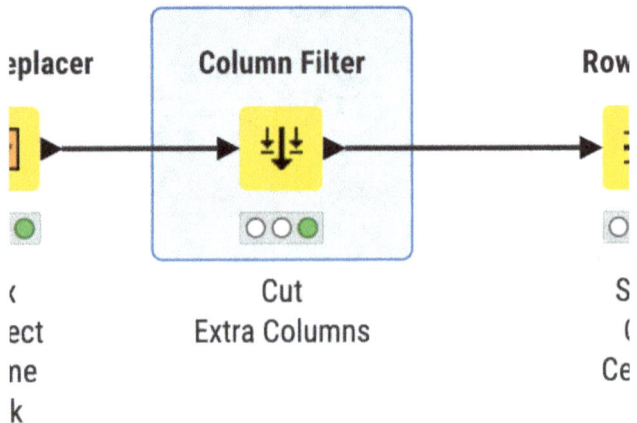

Figure 9-10. Let's Get Choosy

The dialog for the cut is pretty simple – exclusions on the left, inclusions on the right (Figure 9-11).

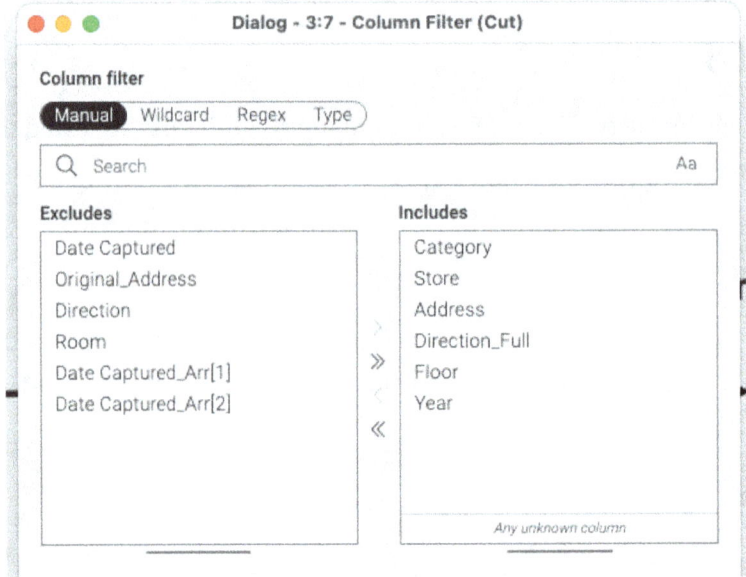

Figure 9-11. Pick What Matters Horizontally

CHAPTER 9   EXAMPLES – PIPELINES

The goal of reducing the data vertically is to help focus on what is truly important for the rest of the analysis. At this point, I could pare the data down to its bare essentials, but I want to leave a little overall context in case something unexpected arises. It's a balancing act of availability and attention span. For example, I've decided to exclude the original "Direction" column and include the more verbose "Direction_Full" – this will make the final analysis a little richer and easier to understand.

> Slice

Since I'm interested in the four directions, I will ignore a new mall section called "Center." There's no analog for this in prior years, so it will be hard to match across these areas; Figure 9-12.

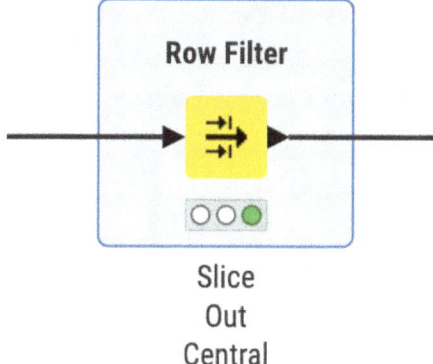

Figure 9-12. Eliminate a New Area to Avoid Duplication

167

CHAPTER 9   EXAMPLES – PIPELINES

Configuring it is as simple as the prior dialog; Figure 9-13.

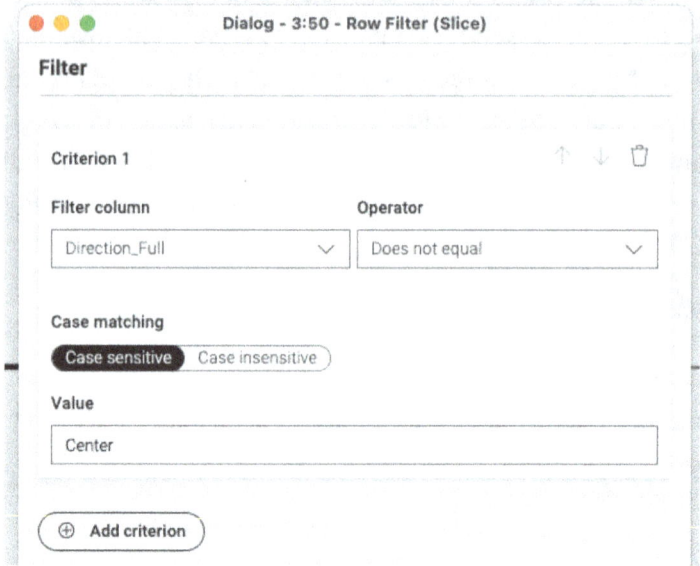

Figure 9-13. Kick Out the "Center" Section

\ Sort

Since the data has opportunities for improvement, I will use the Sorter to organize the stores by name, ascending (Figure 9-14), with the results as seen in Figure 9-15.

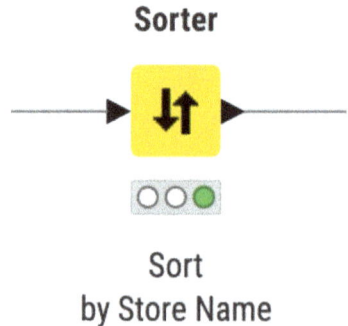

Figure 9-14. Sort Node

CHAPTER 9   EXAMPLES – PIPELINES

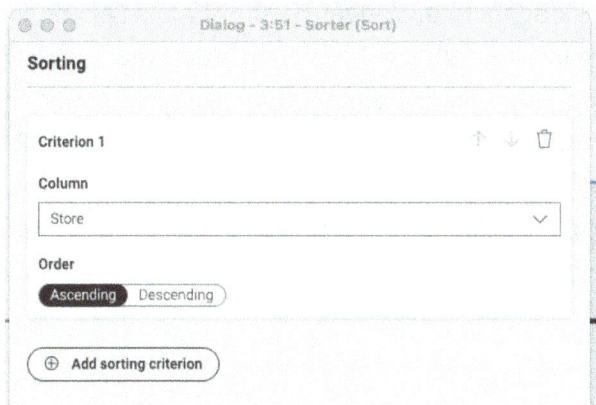

Figure 9-15.   Sorting by Store Number

Build

Now, it's time to build something out of all the data. In my KNIME project, I've processed all four time periods and joined them together where the store name and address match, as shown in Figure 9-16.

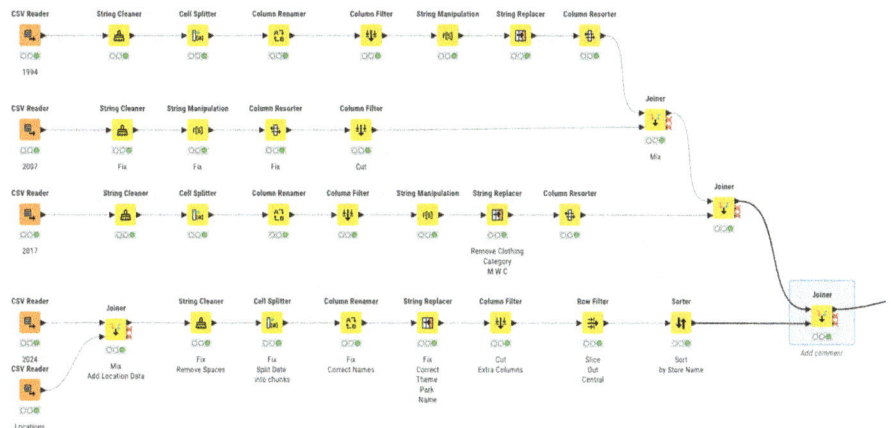

Figure 9-16.   Putting All of the Pieces Together

169

CHAPTER 9 EXAMPLES – PIPELINES

It may look odd that I'm bringing in the context Locations data as a mix for the 2024 year. It wouldn't be helpful for stores that only existed in the earlier decades. However, my goal for the analysis is to identify locations that remain consistent decade after decade, and as a result, the Locations data will apply to all of them. For simplicity, I would probably refactor this later on to be executed after all the decades have been merged.

This results in a table of stores and addresses that remain the same decade over decade; Figure 9-17.

Store String	Address String	Direction_Full String	Floor Number (Integer)
Alpaca Connection	E367	East	3
Ann Taylor	S218	South	2
Barnes & Noble	E118	East	1
Bath & Body Works	N172	North	1
Caribou Coffee	S380	South	3
Cinnabon	E180	East	1
Express	W204	West	2
Helzberg Diamonds	W154	West	1
Hooters	E404	East	4
Johnny Rockets	S370	South	3
Love From Minnesota	W380	West	3
Minnesot-ah!	E157	East	1
Nail Trix	E363	East	3
Nordstrom Rack	W324	West	3
Panda Express	S374	South	3
Perfumania	N219	North	2
Rocky Mountain Chocolate Factory	E128	East	1
Sbarro	S390	South	3
Sox Appeal	W391	West	3
Street Corner News	E122	East	1
Twin City Grill	N130	North	1
Zales	E240	East	2

Figure 9-17. *The Results – 22 Stores*

The original question, which concerns stores opened in 1994 and operating at the exact location and name, is here, but could be summarized into a more useful metric. I want to summarize by direction and floor to see if there's a pattern by location. We can only estimate directional results since the sample size is small, representing less than 5% of the original listing.

CHAPTER 9   EXAMPLES – PIPELINES

] Box

Grouping and sizing in KNIME are done in one step. I can do an entire pivot on the data, including the two groups, in one shot, but I find it easier to visualize when broken out into chunks.

To start, I'll drop in a GroupBy node on the surface; Figure 9-18.

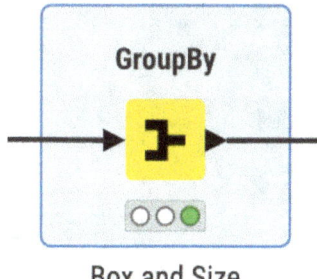

Box and Size

Figure 9-18. *Two for the Price of One*

Configuring the group is done by field selection, as shown in Figure 9-19.

CHAPTER 9   EXAMPLES – PIPELINES

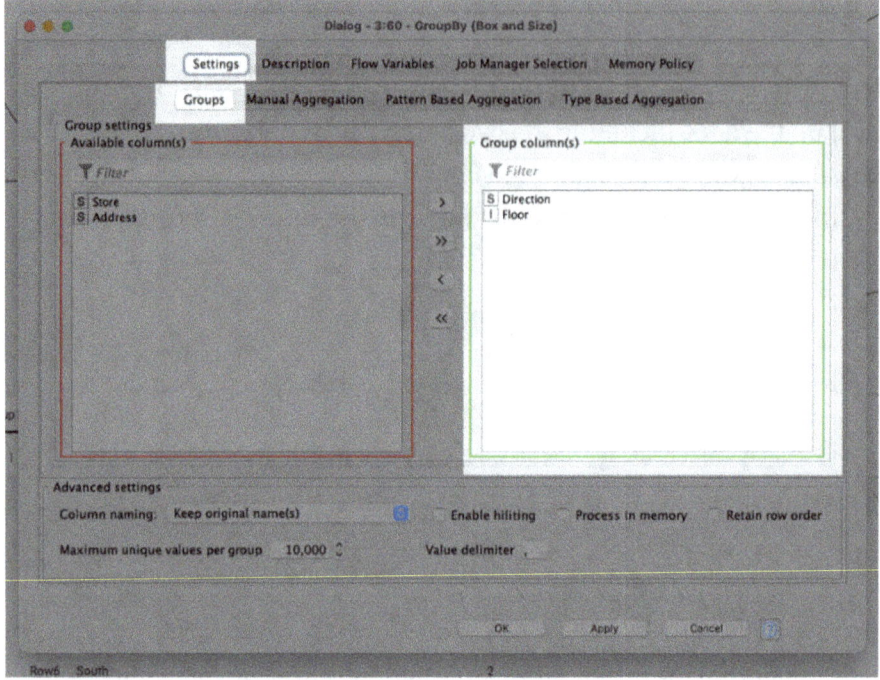

Figure 9-19. *Field Selection – Direction and Floor*

Size

I will do a manual aggregation, simply counting the number of stores; Figure 9-20.

CHAPTER 9   EXAMPLES – PIPELINES

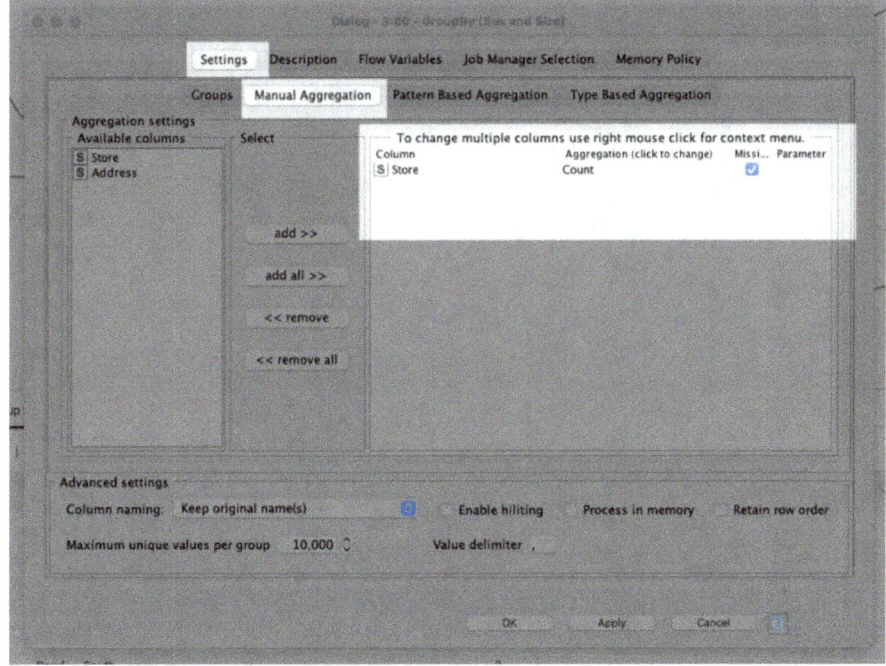

Figure 9-20. *Simple Counts by Store*

The preview for this action shows a simple table with direction, floor, and count of stores; Figure 9-21.

CHAPTER 9   EXAMPLES – PIPELINES

Rows: 11 | Columns: 3

#	RowID	Direction String	Floor Number (integer)	Store Number (integer)
1	Row0	East	1	5
2	Row1	East	2	1
3	Row2	East	3	2
4	Row3	East	4	1
5	Row4	North	1	2
6	Row5	North	2	1
7	Row6	South	2	1
8	Row7	South	3	4
9	Row8	West	1	1
10	Row9	West	2	1
11	Row10	West	3	3

Figure 9-21. Clean Result

= Ship

It's time to create a visual I can share with the world. For this one, I want to make a bubble chart that shows the four cardinal directions on the bottom, counts the floors on the left axis, and defines the size by the number of stores. It's a little complicated but simplifies the data into something readily consumable.

The out-of-the-box bubble chart doesn't quite do what I wanted, so I'm using the community Bubble Chart tool using Plotly; Figure 9-22.

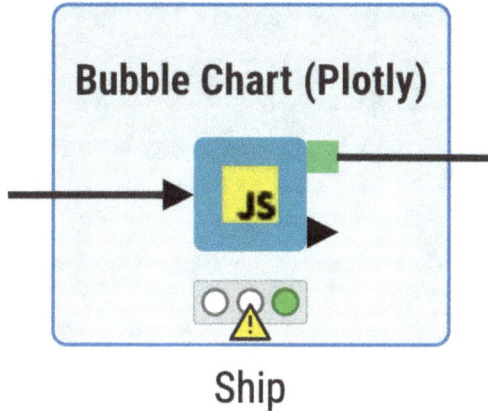

Figure 9-22. *Custom Node for the Bubble Chart*

I want to place the values, so I will use the dialog in Figure 9-23.

CHAPTER 9   EXAMPLES – PIPELINES

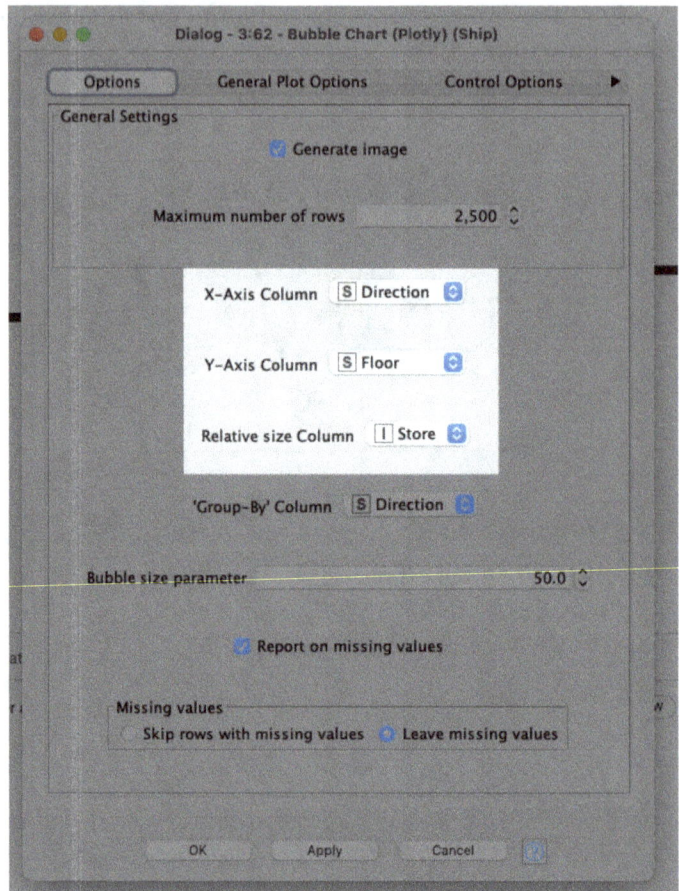

Figure 9-23. Selecting the Corresponding Value for the Axis or Size

This node creates an image that has to be consumed by an "Image View" node; Figure 9-24.

CHAPTER 9   EXAMPLES – PIPELINES

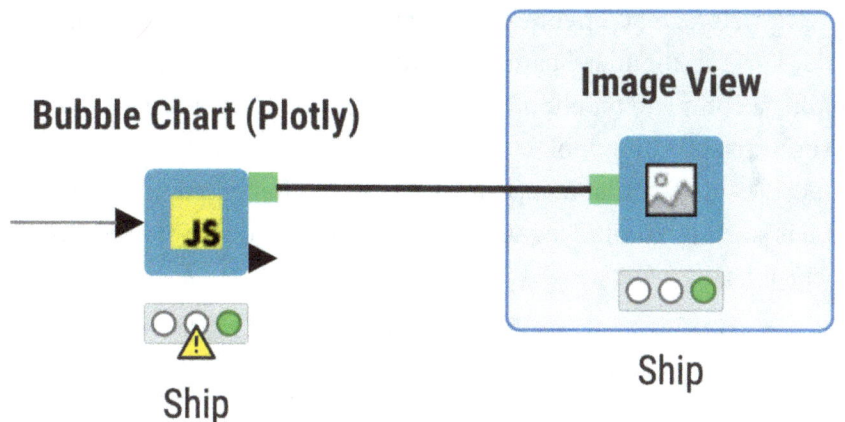

Figure 9-24. *See the Final Results*

It looks like Figure 9-25.

Figure 9-25. *The Big Picture*

177

CHAPTER 9   EXAMPLES – PIPELINES

KNIME is a great pipeline tool, but it lacks power in the visualization world. This rudimentary scatterplot is a great diagnostic, but if I were building a complete pipeline, there would be a second phase with a more powerful visualization tool.

Besides the fact that it appears this mall has fractional floors, the graph is starting to come together. I can pull out a couple of interesting conclusions:

- Most stores are on the first floor on the east side.
- The south and west sides have more stores on the third floor.
- There's some continuity on the fourth floor, East Side. That's where the theater is, so that makes some sense.
- There's a surprising gap on the first floor's south side.
- There was intentionally no coverage for the anchor stores – only two survived – or the many carts sprinkled across the mall hallways.

In a mall with over 500 stores, there is surprisingly little continuity over 30 years. In some cases, stores changed their names multiple times, while in others, they moved to different locations within the mall, depending on traffic patterns.

The Mall of America has intentionally repositioned itself as more of a tourist attraction than a traditional shopping mall. Where we once went to buy school clothes, we now bring out-of-town visitors to marvel at the spectacle. In this era of declining malls, this demonstrates the flexibility and creativity of the owners.

CHAPTER 9   EXAMPLES – PIPELINES

Summary

In this chapter, I used a Pipeline tool to process data from multiple sources and stitch them together in joins to create a final dataset. The overall process is very visual and relatively easy to get the big picture. Putting this into Data Flow Map terms, it looked roughly like:

Motion	Symbol	Action	Goal
Source	O	Get	Pulled in four similar store listing datasets and merged them together
	+	Mix	Added more detailed descriptions of addresses
	X	Fix	Extracted year from date, relabeled columns, and cleaned text
Focus	V	Cut	Trimmed to category, store, address, direction, floor, and year
	>	Slice	Removed the newer central section from locations
	\	Sort	Ordered by store name for readability
Build	]	Box	Grouped by direction and floor
	#	Size	Counted number of stores
	=	Ship	Built a rudimentary bubble chart

Pipeline metaphor apps are powerful tools for connecting data from multiple sources and utilizing various nodes to clean, combine, and summarize this information into valuable insights. The primary drawbacks include a tendency towards spaghetti workflows that complicate understanding how all the pieces fit together, processing power constrained by your computer's CPU, disk, and bandwidth, and, in some cases, exorbitant licensing fees.

CHAPTER 9   EXAMPLES – PIPELINES

Applying the Data Flow Map Framework to a powerful pipeline processor can make it easier to tell the story about how the data was sourced, what decisions were made and along the way, make something amazing and new, as seen in Figure 9-26.

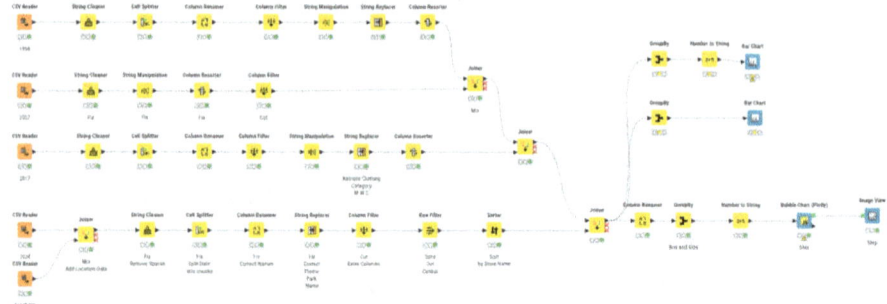

Figure 9-26. *The Entire Pipeline with Multiple Results*

In the next chapter, I cover the analog side of analytics.

CHAPTER 10

Analog – Pen and Paper

This book focuses on the digital side of analytics. How can you quickly understand how a data flow maps to the three key actions – source, focus, and build? Each of the previous chapters went into depth from the perspective of a specific methodology for sourcing data, focusing on the important, and then creating new conclusions.

The final chapter will end the book with something a little different. Let's take the Data Flow Map Framework into your brain as you process what someone says and map the results on paper.

No technology is necessary; however, I can utilize various online drawing and mapping tools. This method offers distinct advantages:

- **Quick Boot Time and Easy Install**: Pick up a pencil or pen; find a piece of paper.
- **Automatic Saves**: Each mark on the paper will be saved until the paper is thrown away.
- **Privacy Is Easy**: Don't share the piece of paper.
- **Collaboration is difficult**; you'll have to huddle around a table and share your pen, but you'll figure it out.

CHAPTER 10   ANALOG – PEN AND PAPER

Applying the Data Flow Map process effectively analyzes code or other existing analytics, breaking it into manageable parts for better understanding and safer improvements.

Do you have a large process to map out? Grab a bigger piece of paper. Your local office supply store should have larger sheets of paper and better pens. Try to avoid overspending.

Are you trying to track someone presenting in a conference room? Keep your notes brief and be prepared for the inevitable questions from your neighbor about the hieroglyphics you're jotting down. Don't hesitate to recommend this book. It's beneficial for the analytic karma in the world.

Finally, I'll include examples of my hand-drawn notes for converting the recipe into a Data Flow Map for two reasons. First, I want you to realize that it doesn't matter how legible your handwriting is, as long as you follow the basic rough symbols. Second, you don't need to use colorful markers, although doing so can help you grasp a process more quickly. Lastly, crossing things out and trying again on a different sheet of paper is perfectly fine. Paper is relatively inexpensive, but your understanding of a process is priceless.

Example Data

My example for this will be mapping out a recipe, specifically a very simple, classic, and ubiquitous recipe for French bread. It's fine if you're not comfortable making bread.

The recipe I'm using is based on Julia Child's, adapted from generations of French boulangeries that made this bread for their local customers. It has the advantage of requiring only a few ingredients. I will simplify the recipe even more for this chapter and my book.

Recipes, by their nature, typically involve sourcing ingredients, focusing on techniques, and refining the process multiple times until you achieve the final delicious product that you can use to impress your friends and family. This DFM will also be a bit unorthodox, but that will be true of any process.

As a reminder, I'll provide the hand-drawn version of the Data Flow Map in Figure 10-1, detailing the three modes and the nine actions.

Figure 10-1. *The Data Flow Map, Informally*

CHAPTER 10   ANALOG – PEN AND PAPER

French Bread – The Data Flow Map
@ Proofing Yeast

This is the beginning of all breads, where you determine whether the yeast you're using is viable. See Figure 10-2.

Figure 10-2. Getting Started

@ Create Dough

Now that the yeast is ready to rock and roll, let's combine it with the flour, as shown in Figure 10-3.

CHAPTER 10   ANALOG – PEN AND PAPER

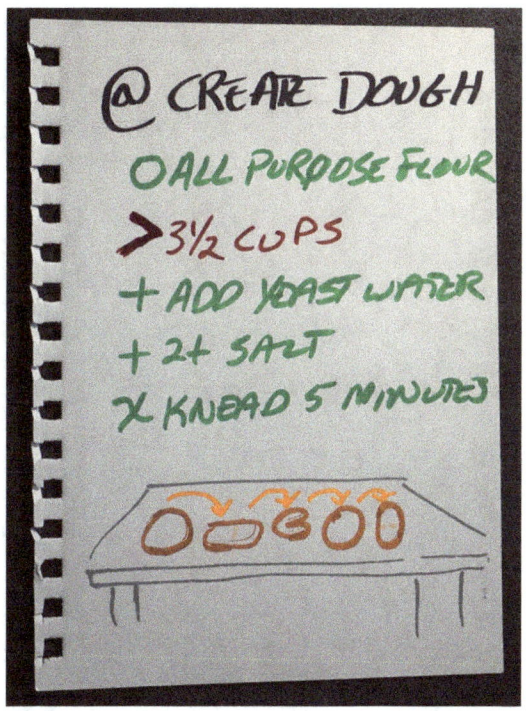

Figure 10-3. Yeast, Plus Flour, and Salt, Then Kneading

@ Rising Dough

Patience, patience. This is where the miracle of growing yeast arrives; Figure 10-4.

Figure 10-4. Waiting, Waiting for Double Time

@ Shaping

It's been a blob until this point. Now, we create the perfect baguette shape; Figure 10-5.

CHAPTER 10 ANALOG – PEN AND PAPER

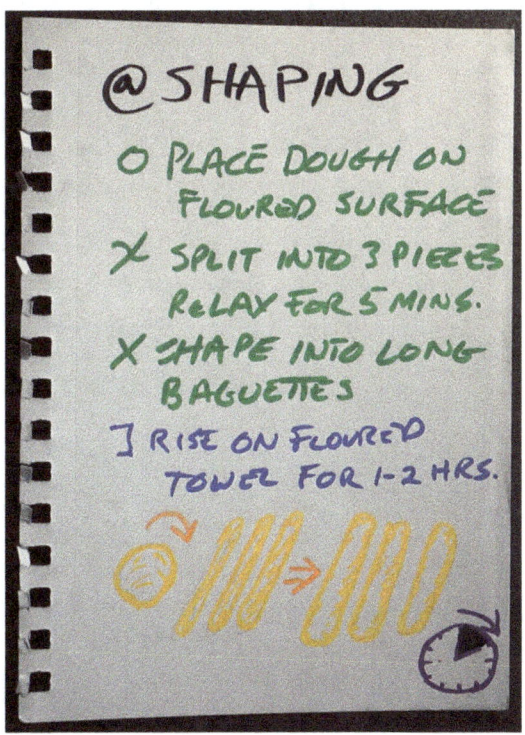

Figure 10-5. Shaping and Yet More Rising

@ Baking

A little water and a lot of heat, and the kitchen smells fantastic, as shown in Figure 10-6.

CHAPTER 10   ANALOG – PEN AND PAPER

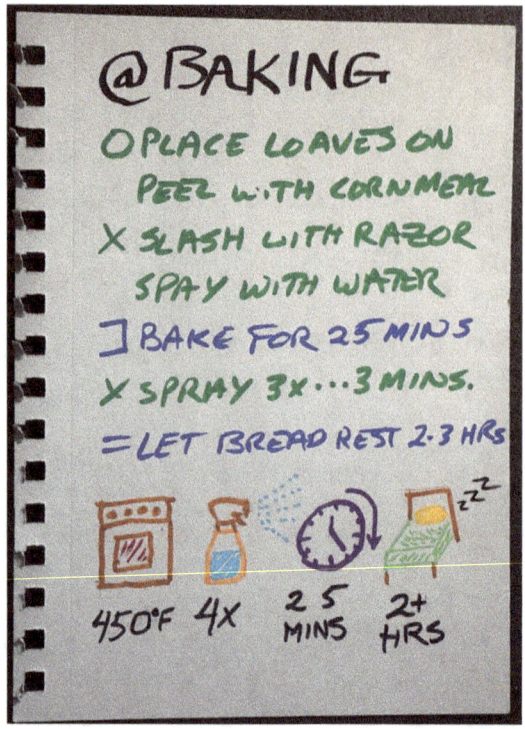

Figure 10-6. The Final Delivery, Yet You Still Wait

Summary

Making sense of the world through data and analytics has historically been a digital process. However, roll back the clock a century or so, computers were people who performed calculations on a page, using, as you might expect, pen and paper.

CHAPTER 10   ANALOG – PEN AND PAPER

Baking bread is an iterative process going through several checkpoints. I'll block out the Data Flow Map accordingly into smaller tables, beginning with Table 10-1.

- Table 10-1: Getting started
- Table 10-2: Pulling together ingredients
- Table 10-3: Building something new
- Table 10-4: Creating shapes
- Table 10-5: Grand Finale

Table 10-1. Start at the Beginning

@	**Proof the yeast**
0	Source Water
>	Only 1 ½ cups
X	Fill Small Bow
+	Add in 2 T Yeast
]	Wait 5 minutes for bubbles

Table 10-2. Build the Main Scene

@	**Create the Dough**
0	All Purpose Flour
>	Only 3 ½ Cups
+	Add Yeast Mixture
+	Add 2 t Salt
X	Knead 5 Minutes

189

CHAPTER 10   ANALOG – PEN AND PAPER

Table 10-3. *Let Things Build*

@	**Rising Dough**
0	Kneaded Dough
]	Place in a bowl
#	3.5 x size over ~ 3 Hours
]	Deflate and put back in bowl
#	3x size ~ over 1-2 Hours

Table 10-4. *What's It Going to Look Like?*

@	**Shaping**
0	Put risen dough on floured surface
X	Split into 3 pieces, Relax for 5 minutes
X	Shape into long baguettes
]	Rise on floured towel for 1–2 hours

Table 10-5. *We're Almost Done Here*

@	**Baking**
0	Place loaves on cornmeal on peel
X	Slash with razor, spray with water
]	Bake for 25 minutes
X	Spray 3x every 3 minutes
=	Let baked bread rest 2–3 hours

It's very common to use computers of any size and portability to capture someone's thoughts and document them in video or transcription. However, if you find yourself in a situation where you want to converse with someone without a screen between you, or if you're trying to identify the larger patterns of thought, you can use the analog version of the Data Flow Map to quickly illustrate how someone moves from point A to point B. The person might be using software you've never encountered before, or they might be teaching you a new process, such as baking bread.

Julia Child was a master at clarity – she could break down a complex and very French cooking process into its basic elements, tell the story clearly, and apply a little of her own creativity. Every person preparing food, whether a homemaker providing for the family or a famous chef serving hundreds per night at their fine dining establishment, faces the same challenges – understanding the process clearly, communicating effectively with family or staff, and creating something surprising and new.

I hope that the Data Flow Map Framework provides you with the tools to develop new analytics, processing data in a clear and straightforward way that is easy for others to understand and communicate, regardless of the data source or tools used.

See Appendix A for sample data sourcing.

APPENDIX A

Sample Data Sourcing

This book uses various sample datasets from the Internet and my own experiences. All data is intentionally brief, often anonymous, and, most importantly, provided to exercise different tools for various purposes.

Table A-1 lists the datasets and their brief descriptions:

Table A-1. *Sample Datasets*

Context	Datasets	Purpose
4 – Files	Weather Observations History	Prepared CSV File with observations
5 – Databases	Coffee Purchases	Ten years of coffee choices across a selection of coffee shops, along with the shop categories
6 – Code	Flights	Flight tracking data using a Raspberry Pi
7 – Cloud API	Weather Observations Current	Using API calls, pull the current weather and save it to JSON files
8 – Platforms	Minnesota Lake Towns	City descriptors and proximity to Twin Cities metro region
9 – Pipelines	Mall Tenancy History	Four time periods of mall tenant information across 30 years
10 – Bread	French Bread Recipe	Demonstrate capturing a process using Data Flow Map

APPENDIX A   SAMPLE DATA SOURCING

Weather History

Our local weather can experience extremes, and tracking it is straightforward. The API example details this more; however, this example only involves taking hourly observations of the local weather.

This relies on METAR data, which is designed more for pilots flying planes than for the average person walking their dog, but the basic information remains the same. Contextual data helps convert cloud coverage into a percentage that can be played with when mixing data.

The raw data looks like Figure A-1.

Type	Station	Observed_Da	Temp_Celsiu	DewPoint_Ce	Wind_Speed	Wind_Directi	Cover	Visibility	Observation Logg
METAR	KMSP	2022-12-27T	-15.6	-20	5	220	BKN	10	KMSP 270253 2022
METAR	KMSP	2022-12-27T	-15.6	-20	5	220	BKN	10	KMSP 270253 2022
METAR	KMSP	2022-12-27T	-16.1	-21.1	5	200	SCT	10	KMSP 270353 2022
METAR	KMSP	2022-12-27T	-16.1	-20.6	0	0	BKN	10	KMSP 270453 2022
METAR	KMSP	2022-12-27T	-16.1	-20.6	6	180	BKN	10	KMSP 270553 2022
METAR	KMSP	2022-12-27T	-16.1	-20.6	5	170	BKN	10	KMSP 270653 2022
METAR	KMSP	2022-12-27T	-16.1	-20.6	7	150	OVC	10	KMSP 270753 2022
METAR	KMSP	2022-12-27T	-15	-20	7	160	OVC	10	KMSP 270853 2022
METAR	KMSP	2022-12-27T	-13.9	-19.4	7	130	BKN	10	KMSP 270953 2022
METAR	KMSP	2022-12-27T	-13.3	-18.9	9	150	OVC	10	KMSP 271053 2022
METAR	KMSP	2022-12-27T	-12.2	-17.8	9	160	OVC	10	KMSP 271153 2022
METAR	KMSP	2022-12-27T	-11.7	-17.2	9	170	SCT	10	KMSP 271253 2022
METAR	KMSP	2022-12-27T	-11.7	-16.7	9	150	BKN	10	KMSP 271353 2022
METAR	KMSP	2022-12-27T	-9.4	-15.6	13	170	BKN	10	KMSP 271453 2022
METAR	KMSP	2022-12-27T	-8.3	-14.4	8	180	BKN	10	KMSP 271553 2022

Figure A-1. *Sample Weather Data*

Coverage types are much more straightforward; see Figure A-2.

Cover_Type	Cover_Long	Max_Coverage_pct
CLR	Clear	0
FEW	Few Clouds	0.25
SCT	Scattered	0.5
BKN	Broken	0.88
OVC	Overcast	1
OVX	Obscured	

Figure A-2. *Cloud Coverage Conversions*

APPENDIX A   SAMPLE DATA SOURCING

Coffee Purchases

This data is sourced and slightly anonymized from my coffee shopping history over the last ten years. Yes, I do have a habit.

There are two primary tables:

1. Coffee Purchase History – daily shopping data, including the purchase method, date, location name, and purchase amount. See Figure A-3.

Method	Date	Payee	Amount
Checking	1/6/14	Dunn Brothers Coffee	$ 2.09
Checking	1/6/14	Dunn Brothers Coffee	$ 2.09
Checking	1/6/14	Starbucks	$ 10.00
Checking	1/6/14	Starbucks	$ 10.00
Checking	1/6/14	Starbucks	$ 3.43
Checking	1/6/14	Starbucks	$ 1.84
Checking	1/7/14	Dunn Brothers Coffee	$ 2.09
Checking	1/7/14	Starbucks	$ 5.00
Checking	1/8/14	Starbucks	$ 18.89
Checking	1/9/14	Dunn Brothers Coffee	$ 4.64
Checking	1/9/14	Starbucks	$ 3.43

Figure A-3. *Coffee Purchase History*

Store Category – location name and size category. See Figure A-4. Sizes range from:

- **Nationwide**: Starbucks and Target
- **Regional**: Multi-State Coverage
- **Metro**: Just in the local metropolitan area
- **Local**: Single, one-owner shops

APPENDIX A   SAMPLE DATA SOURCING

Store	Category	Location Count
Starbucks	National	40000
Caribou	National	750
Dunn Brothers Coffee	Regional	48
Urban Coffee	Local	1
Sencha Tea Garden	Met	3
Tea Garden	Local	1
Bull Run Coffee	Local	2
Studio 2 Café	Local	1
Target Store	National	1700
Moose And Sadies	Local	1

Figure A-4. *Location Category*

Flight Tracking

Several online services allow you to track an airplane's actual path. The data is sourced from the publicly available transmissions of the airplanes themselves. It can be captured and processed using FlightAware software. In my case, I set up a Raspberry Pi with an antenna readily available online and started capturing the radio messages. Processing the text messages is an exercise for this book.

As the software captures the data, it's stored in a series of historical JSON files that are overwritten every two hours. The directory of files looks a lot like Figure A-5.

APPENDIX A   SAMPLE DATA SOURCING

Figure A-5. *File Listing of Raw Flight Data*

Weather Observations – Current

Unlike the historical weather file, this is the weather as it is right now. Pulling the data from the publicly available METARS servers is free for reasonable use.

I use a script scheduled to run regularly, pulling the data for the stations I'm interested in. Like the entire process documented in the API chapter, it's possible to do this entirely in command-line shell code. It is reasonably efficient for a small computer like a Raspberry Pi.

The data is set up with a flexible JSON format that makes it easy to parse.

Minnesota Lake Towns

Geographical data is available on the State of Minnesota data sites. They have a wealth of information about the State's business and geography.

The distance information is gathered using latitude and longitude data, which are readily available on Wikipedia.

Calculating the distance between the towns and the center point of the Twin Cities is done in code using Python. The code looks approximately like Listing A-1.

197

APPENDIX A   SAMPLE DATA SOURCING

Listing A-1. Processing Distance

```
import math

def equirectangular_distance(lat1, lon1, lat2, lon2,
radius=6371):
    """
    Calculate distance between two points using the
    equirectangular approximation.

    Parameters:
        lat1, lon1 -- Latitude and Longitude of point 1 (in
        degrees)
        lat2, lon2 -- Latitude and Longitude of point 2 (in
        degrees)
        radius -- Radius of the Earth (default 6371 km; use
        3959 for miles)

    Returns:
        Distance in the same unit as the radius (km or miles)
    """
    # Convert degrees to radians
    lat1_rad = math.radians(lat1)
    lon1_rad = math.radians(lon1)
    lat2_rad = math.radians(lat2)
    lon2_rad = math.radians(lon2)

    # Compute deltas
    x = (lon2_rad - lon1_rad) * math.cos((lat1_rad + lat2_
    rad) / 2)
    y = lat2_rad - lat1_rad

    # Compute distance
    distance = radius * math.sqrt(x*x + y*y)
    return distance
```

APPENDIX A   SAMPLE DATA SOURCING

Mall Tenancy History

The Mall of America has listed its tenants on the Internet for many years. Using the Wayback Machine (`https://archive.org/web/`), you can pull earlier copies of the website. In this case, I could return to 1997, the very dawn of the public Internet.

The website has changed a lot over time, but that was mostly a matter of applying different Python scripts to line up the data roughly in the order presented in the book. For the first listing of the Mall's tenants, I got a PDF of the original mall map, which I processed using GenAI.

The data is incomplete, and different opinions on what's important over the years have led to some wildly different results. But for the chapter, it's enough to give us a starting point.

French Bread Recipe

These recipes are about as old as time; they're really just variations on a theme now. Julia Child took a fairly "standard" recipe, if there is one, Americanized it, and then presented it on public television.

If you want the original, you can buy Julia's book yourself, or if you search the internet, you'll find thousands of variations on her theme. If you'd like to see her in action, check out this video: `https://www.dailymotion.com/video/x2gtlad`.

The version I present is even more simplified, but if followed, it will produce a respectable set of French bread loaves. If you're curious about doing it "correctly," search for "French Bread The French Chef Season 7" and get hers along with a host of other takes on the same theme.

APPENDIX A   SAMPLE DATA SOURCING

Templates

It's helpful to see how the Data Flow Map might surface in the wild in the context of the tools that you will use to document and execute code. The following are example templates tailored for the respective environments.

Markdown

Now one of the most common formats for text documentation, it's very useful for documenting Data Flows. The DFM can be adapted in table form easily, see Listing A-2. The backticks around the action symbol will format it neatly in fixed font when run through Github or static site generators.

Listing A-2. Markdown Tables

```
`@` | Checkpoint Tag
--- | ---
`O` | Get
`+` | Mix
`X` | Fix
`V` | Cut
`>` | Slice
`\` | Sort
`]` | Box
`#` | Size
`=` | Ship
```

When rendered in Markdown, this will look something like Figure A-6.

APPENDIX A   SAMPLE DATA SOURCING

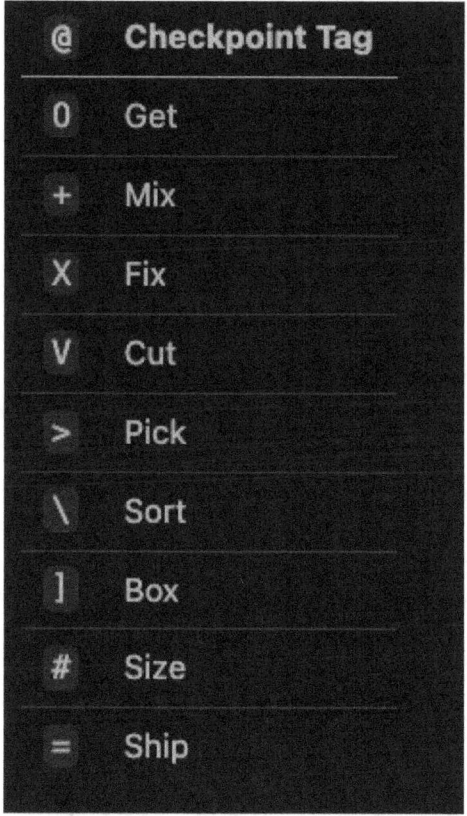

Figure A-6. *Rendered Markdown*

SQL

If you have a particularly complex set of SQL, sometimes it can be helpful to insert the DFM throughout or just at the beginning. Traditionally, most SQL processors will ignore a triple dash as comments.

It'll be easier for you later to understand your code, see Listing A-3.

APPENDIX A   SAMPLE DATA SOURCING

Listing A-3. SQL Commentary DFM

```
--- @ Checkpoint Tag
--- -
--- O Get
--- + Mix
--- X Fix
--- V Cut
--- > Slice
--- \ Sort
--- ] Box
--- # Size
--- = Ship
```

Code

In this example, I'll focus on Python, since that's probably the most common language used with Data; however, as long as you follow your preferred language's commenting scheme, you should be fine.

The big challenge is the hashtag *is* a comment indicator in Python. For the sake of clarity, I'll use the triple tick mark to block out the code, see Listing A-4.

Listing A-4. Python Commentary

```
'''
@ Checkpoint Tag
---
O Get
+ Mix
X Fix
V Cut
```

APPENDIX A   SAMPLE DATA SOURCING

```
>  Slice
\  Sort
]  Box
#  Size
=  Ship
'''
```

A bit of Python code will then look like Figure A-7.

```python
import os
import sys

'''
@ Checkpoint Tag
___
O Get
+ Mix
X Fix
V Cut
> Slice
] Box
# Size
= Ship
'''

def get_file_path(filename):
    """
    Get the absolute path of a file in the current directory.
    """
    current_dir = os.path.dirname(os.path.abspath(__file__))
    return os.path.join(current_dir, filename)
```

Figure A-7. *Data Flow Map in Code*

APPENDIX A   SAMPLE DATA SOURCING

Handwritten

While I've demonstrated the form in Chapter 10, I'll provide a couple of different handwritten versions for your use. For example, Figure A-8, illustrates the vertical version, and Figure A-9, the more condensed horizontal version.

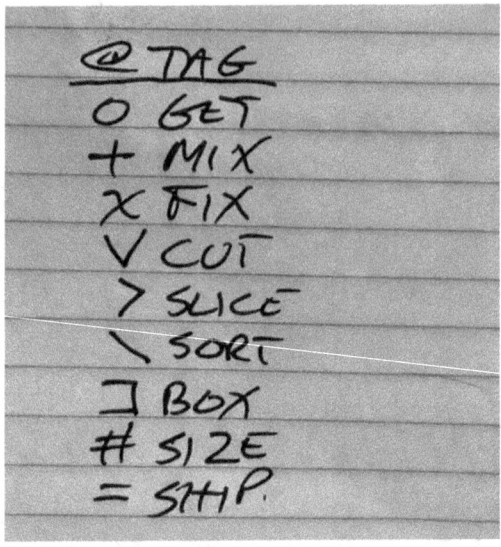

Figure A-8. Vertical

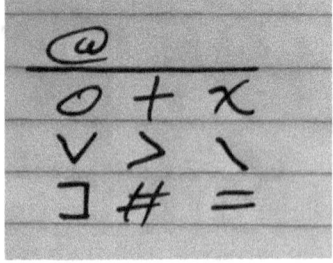

Figure A-9. Condensed

GPSR Compliance
The European Union's (EU) General Product Safety Regulation (GPSR) is a set of rules that requires consumer products to be safe and our obligations to ensure this.

If you have any concerns about our products, you can contact us on

ProductSafety@springernature.com

In case Publisher is established outside the EU, the EU authorized representative is:

Springer Nature Customer Service Center GmbH
Europaplatz 3
69115 Heidelberg, Germany

www.ingramcontent.com/pod-product-compliance
Lightning Source LLC
LaVergne TN
LVHW020411070526
838199LV00054B/3581